THE COMMONWEALTH AND INTERNATIONAL LIBRARY
Joint Chairmen of the Honorary Editorial Advisory Board
SIR ROBERT ROBINSON, O.M., F.R.S., LONDON
DEAN ATHELSTAN SPILHAUS, MINNESOTA

PERGAMON OXFORD GEOGRAPHIES
General Editor: W. B. FISHER

RURAL GEOGRAPHY

AN INTRODUCTORY SURVEY

RURAL GEOGRAPHY

AN INTRODUCTORY SURVEY

BY

HUGH D. CLOUT

PERGAMON PRESS

OXFORD . NEW YORK . TORONTO
SYDNEY . BRAUNSCHWEIG

PERGAMON PRESS LTD., Headington Hill Hall, Oxford
PERGAMON PRESS INC., Maxwell House, Fairview Park, Elmsford, New York 10523
PERGAMON OF CANADA LTD., 207 Queen's Quay West, Toronto 1
PERGAMON PRESS (AUST.) PTY. LTD., Rushcutters Bay, N.S.W. 2011, Australia
VIEWEG & SON GmbH, Burgplatz 1, Braunschweig

First edition 1972

Library of Congress Catalog Card No. 72–87827

Printed in Great Britain by A. Wheaton & Co., Exeter

08 017042 0 Flexicover
08 017041 2 Hard cover

CONTENTS

LIST OF FIGURES

LIST OF TABLES

PREFACE

OVER recent years important social, economic, and land-use changes have taken place in rural areas in the developed world, but these have largely passed unrecorded by the authors of textbooks on human geography. The chapters which follow represent an attempt to draw attention to a selection of key issues that have been investigated by geographers, sociologists, and other social scientists. It has not been my intention to write a "complete" rural geography tracing the historical evolution of rural problems on a worldwide scale. My research experience of rural matters is limited to a selection of themes in parts of western Europe. The pages which follow reflect that fact. Each chapter is supplemented by a fairly extensive bibliography, allowing the text to be used at two levels: first, as a review of selected rural themes, and, second, as a starting point for more detailed investigation by students with access to well-stocked libraries.

I am indebted to my teachers at University College London—J. T. Coppock, H. C. Darby, J. H. Johnson, H. C. Prince, and D. Thomas—who introduced me as an undergraduate to a wide range of historical and contemporary rural problems in western Europe. I have also been encouraged by the published works of H. E. Bracey, R. E. Pahl, and G. P. Wibberley. My thanks are extended to my colleague R. J. C. Munton for his willingness to discuss rural issues and for his permission to use illustrations drawn from joint research. André Fel has taught me much about the practical problems involved in rural management in backward areas.

Professor W. R. Mead allowed me to develop teaching in rural geography at University College London. I wish to thank him and also the groups of students who have contributed to discussions and seminars on rural themes in the C20 and C30 courses over the last few years. Their own ideas, criticisms of my ideas, and general willingness to investigate topics in depth have been stimulating to a far greater degree than I suspect they think. I wish to record my thanks to my parents for acting as lay readers of the manuscript, and to members of the Cartographic Unit of UCL (especially Margaret Thomas) for their skilful preparation of the illustrations.

HUGH D. CLOUT

ACKNOWLEDGEMENTS

The author and the publishers thank all those who have given permission to use copyright material and in particular thank the following who have allowed adaptations of illustrations from other publications to be used in this book:

Lunds Universitet, for the adaptation of an illustration from *Studies in rural/urban interaction* by E. Kant, from Lund Studies in Geography, Series B, 1951; The Agricultural University of Wageningen for an adaptation of map in "Outdoor Recreation in Sweden" by C. E. Norrbom from *Sociologica Ruralis*; Librairie Armand Colin for material adapted from *La Siderugie française* No. 374 (C. Precheur) and *Les Paysages agraires* Series 329 (A. Meynier); Hermann Paris for a simplified version of a map from Jean Labasse, *L'organisation de l'espace*, Paris, 1971; The Controller of Her Majesty's Stationery Office for a simplified version of a map from *A Century of Agricultural Statistics, Great Britain*, 1866–1966; Thomas Nelson & Sons Ltd. for a version of a map by R. Lawton in *The British Isles* (J. Wreford Watson and J. B. Sissons), 1964; PUDOC and the Elsevier Publishing Company for a version of Map 3 in *Agro-Ecological Atlas of Cereal Growing in Europe*, vol. i (Brockhuizen); Oriel Press Ltd. for adaptations of maps from "The Remote Countryside: A Plan for Contraction" (R. J. Green) from *Planning Outlook*, 1966; The Regional Studies Association for a map by R. Lawton in "The Journey to Work in Great Britain" from *Regional Studies*, 1968; *The Scottish Geographical Magazine* for a simplified version of a map by I. Robertson from "The Occupational Structure and Distribution of Rural Population in England and Wales" from volume 17, 1961; The Countryside Commission for a simplified version of a map from *The Countryside Commission*; The Highlands and Islands Development Board for their kind co-operation in providing the map of the Board schemes; M. Poitrineau for the use of an illustration in *Demographie Historique* published by Librairie Armand Colin; Professor André Fel for the use of a simplified version of a drawing by C. Mignon published in *Revue d'Auvergne*, 1971.

While every attempt has been made to trace copyright holders this has not been possible in all cases and the author and the publisher would be glad to hear from copyright holders of any other material used in *Rural Geography*.

CHAPTER 1

RURAL GEOGRAPHY: AN OVERVIEW

THE SCOPE OF RURAL GEOGRAPHY

From being at the core of studies in human geography prior to World War II, the countryside, as a field of geographical investigation, has been relegated to an inferior position. Such a decline in interest contrasts with the rapid and sophisticated growth of research and expertise in aspects of urban geography and, of course, reflects the impact of the worldwide process of urbanization. It is the aim of the chapters which follow to show that important social, economic, and land-use changes are taking place in the countryside which merit rather more attention from geographers than they currently receive.

The term "rural geography" might be used to describe several realms of geographical knowledge. It might be applied to the economic geography of agricultural production. This was certainly the interpretation adopted by Pierre George in his *Précis de géographie rurale* (1963) which presented a worldwide review of "fundamental features of rural life and the objectives and difficulties of agricultural production in different physical, economic and social environments" (p. 1). Such an approach will not be followed in the present book which is not intended to be a discussion of economic geography *per se*. Neither is "rural geography" taken to be synonymous with "rural settlement" which involves investigations of the patterns, origins, and functions of settlements. In some aspects such studies might be seen as closely allied to historical geography, and, in others, as a branch of quantitative investigation. For the purposes of the discussion which follows, rural geography may be defined as the study of recent social, economic, land-use, and spatial changes that have taken place in less-densely populated areas which are commonly recognized by virtue of their visual components as "countryside".

Thus one could follow the definition of "rural" proposed by G. P. Wibberley (1972): "The word describes those parts of a country which show unmistakable signs of being dominated by extensive uses of land, either at the present time or in the immediate past. It is important to emphasise that these extensive uses might have had a domination over an area which has now gone because this allows us to look at settlements which to the eye still appear to be rural but which, in practice, are merely an extension of the city resulting from the development of the commuter train and the private motor car" (p. 2).

One might argue that in a rapidly urbanizing world the countryside is of

1

decreasing significance and does not merit special investigation. It is certainly true that more people are living in cities, but such a trend has both formal and functional implications for the surrounding countryside. In formal terms, land is abstracted from agricultural and other rural uses and is covered with bricks, mortar, and tarmac. This is illustrated by Fig. 1.1 which attempts to provide a simplified summary of some of the themes that are of importance in rural geography. The right-hand column lists nine major demands or pressures on rural land. Some of these, such as food production, water gathering, and recreation, would remain as "rural" users of land. Others, including housing and manufacturing industry, are less obviously rural. If occurring over a broad area, such uses might involve the conversion of sections of the countryside into townscape. These demands on rural land form areas of actual or potential conflict which need to be resolved by efficient land-use planning. An illustration of the passage of agricultural land to non-agricultural use will be presented in the British context later in this introductory chapter.

Important functional changes are occurring in the countryside as agriculture and other traditional rural activities release workers from local employment. Depopulation and a process of ruralization may be recognized whereby settlements in remote areas contain not only smaller total populations but also a predominance of agricultural workers, whereas in the past they had supported larger total numbers which included craftsmen and service workers as well as farmers and farmworkers. Just the reverse process is taking place in other parts of the countryside which, after having gone through a phase of depopulation, are now being invaded by highly mobile city dwellers for purposes of residence, recreation, and retirement. Repopulation and urbanization of the countryside are taking place and are manifesting themselves in the form of "metropolitan" or commuter villages, second homes, and all the trappings of the recreation industry. Not surprisingly, significant social changes have resulted in these invaded areas which, in visual and spatial terms, are still countryside.

From a sociological point of view it is possible to argue that rural people and rural communities may no longer be identified in developed parts of the world, such as Great Britain, western Europe, and the United States, where all—or virtually all—inhabitants have now acquired urban-derived aspirations, enjoy urban life styles, and form parts of national systems of mass culture. The details of this type of argument will be presented in later chapters. Such issues may shake the underpinnings of rural sociology as a separate branch of social science, but they do not weaken the geographer's concern for the countryside, which forms a different—indeed compensatory—environment from the city. Geographers are, nevertheless, aware of the important diffusion of urban influences into the countryside and take account of such a wide-reaching process in their analyses of rural phenomena.

Important changes in man–land ratios have resulted in the countryside from both depopulation and repopulation. They point to the need for a rational management of rural resources to cater for a reduced number of permanent rural residents in remote country areas, for seasonal invasions by visitors, and for the repopulation of peri-urban areas which form parts of the dispersed city. Farms,

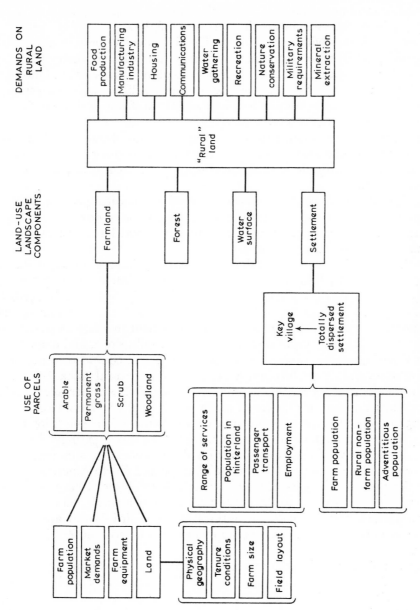

FIG. 1.1. Some components of rural geography.

fields, and other agricultural structures need to be remodelled to meet the requirements of modernized farming. A wide range of other rural structures, from settlement patterns to systems of transport and traditional concepts of land use, need to
be recast to meet new demands. Particular emphasis will be placed on aspects of
change in the countryside in the chapters which follow.

Three key themes will be presented. The first involves outmigration from the
countryside, leading to depopulation which has affected many parts of the
developed world during and after their periods of industrialization. Outmigration
is viewed at a variety of scales, working from generalized economic explanations
through a consideration of more localized social reasons to eventually focus down
on the outmigrant as a decision-maker.

The second theme involves the recent repopulation of some rural areas in the
post-war age of increased personal mobility as car-owning urbanites choose to
move into the countryside for residence—on a permanent or a part-time basis—
and for recreation.

The third group of chapters discusses selected systematic aspects of planned and
unplanned change in the countryside. Issues as diverse as settlement rationalization, the conversion of agricultural land to forest, and landscape evaluation for
planning purposes will be presented.

In reality, all these systematic themes coexist and interact in particular spatial
or regional contexts. It is not, however, the author's intention to present a series
of regional case studies. These may be investigated elsewhere. Nevertheless, a
very strong case may be made for the need for the integrated management of
rural areas on a geographic basis. Such management involves an understanding
of spatial variations in physical resources such as soils, slope, and climate, as well
as social and economic issues. The geographer is well equipped to tackle such a
range of issues and should be prepared to play a role in rural management.

The treatment of themes in the chapters which follow is necessarily selective.
It represents the author's choice of issues which might be considered important
rather than any attempt to build up a comprehensive framework for rural geography as a sub-branch of the discipline. Discussion is cast entirely in the context
of the developed world with emphasis on the British Isles, although occasional
excursions will be made to Continental Europe and North America to illustrate
particular themes. Many issues presented at a British national level are, however,
of more general relevance to the developed world as a whole. Rural problems in
the varied and complex cultural settings of the Third World are beyond the
experience of the author and are certainly outside the scope of the chapters which
follow.

Spatial and temporal variations in the use of resources by man form a central
theme for the whole of human geography. The section which follows presents a
brief overview of the abstraction of land from rural uses in Great Britain and
thereby provides both an illustration of the kinds of land-use change that are
experienced in the developed world and also a setting for the various themes that
will be presented later.

CHANGING BALANCE OF LAND USE IN GREAT BRITAIN

During the twentieth century there have been important changes in national patterns of land use as increasingly large areas have been absorbed for residential uses and the attendant trappings of urban civilization such as airports, highways, and industrial sites. In spite of the importance of such changes, relatively little is known of their volume or spatial distribution in most countries of the world. The situation in Great Britain, however, has been studied in detail and will serve by way of illustration.

The agricultural surface of England and Wales continued to expand up to the end of the nineteenth century as scrubland was reclaimed for intensive use. Since then the farmed surface has contracted because of competition for land from urban development and forestry. The area devoted to urban growth was restricted until after 1918, even though the population living in densely built British towns increased rapidly in the nineteenth and early twentieth centuries. Conditions changed in the 1920's when planning control was virtually non-existent and agriculture was in a very depressed state. Building sites were cheap to purchase in the countryside which extending public transport services and rising rates of car-ownership made progressively more accessible. Prosperity revived after the depression, and the annual rate of transfer of farmland to urban uses was greater between 1934 and 1939 than for any other five-year period before or since (Table

TABLE 1. TRANSFERS OF AGRICULTURAL LAND IN ENGLAND AND WALES

	Annual average transfers to . . .		
	Urban development (ha)	Service departments and miscellaneous (ha)	Woodland and forestry (ha)
1927/8–1933/4	18,900	500	?
1934/5–1938/9	24,500	6,100	?
1939/40–1944/5	4,300	41,100	7,200
1945/6–1964/5	15,600	5,000*	7,900

* Represents a reduction in service property.

Source: BEST, R. H. (1968b), Competition for land between rural and urban uses, *Land Use and Resources: Studies in Applied Geography*, Institute of British Geographers Special Publication, **1**, 93.

1). In the first 60 years of the present century the urban surface of England and Wales doubled, rising from 5.4 per cent to 10.8 per cent of the land surface. The urban area of Great Britain increased by 10 per cent each decade through that period but absorbed under 1 per cent of the country's land every 10 years. In 1900, 88.7 per cent of England and Wales (Table 2) had been composed of "rural"

TABLE 2. CHANGING LAND-USE BALANCE OF ENGLAND AND WALES, 1900–60

	Agriculture (%)	Woodland (%)	Urban development (%)	Unaccounted for (%)
1900	83.6	5.1	5.4	5.9
1925	82.9	5.1	6.2	5.8
1935	81.8	5.7	7.6	4.9
1950	80.6	6.4	9.7	3.3
1960	79.3	6.8	10.8	3.1
Great Britain 1960	82.4	7.4	7.9	2.3

Source: WIBBERLEY, G. P. (1967), The pressures on Britain's rural land, in ASHTON, J. and ROBERTS, S. J. (eds.), *Economic Change and Agriculture*, Oliver & Boyd, Edinburgh, p. 157.

land (farmland plus woodland). Sixty years later 86.1 per cent still remained in rural use. In Great Britain 89.8 per cent of the land surface was rural in 1960.

R. H. Best (1968b) explains that it would not be unrealistic "to expect that the total urban area of Great Britain will extend by a further 725,000–770,000 ha from 1960 to the end of the century. A growth of this order would involve an average transfer of farmland to urban use at the rate of 18,000–19,000 ha a year, or rather more than the present average. . . . Even by the year 2000 the total area of urban land will probably not occupy more than 15–16 per cent of the whole land surface, leaving some 84–85 per cent of the country in some form of rural use" (p. 98). Thus the general spatial substance of rural geography will remain, but detailed components of rural landscapes will undergo important changes by the end of the century. Modern fields, factory farms, large machine sheds, and vast hedgeless fields will combine with an increase in the afforested area to provoke significant changes in rural landscapes not only in Great Britain but in all other countries of the developed world. These visual changes will be paralleled by changes in the numbers and social characteristics of rural residents and urban people making use of country resources for short periods. Historic rural depopulation will give way to various types of seasonal or permanent repopulation in many parts of the countryside.

REFERENCES AND FURTHER READING

Valuable background material to a study of rural geography is contained in:

ARVILL, R. (1966) *Man and Environment*, Penguin, Harmondsworth.
BRACEY, H. E. (1970) *People and the Countryside*, Routledge & Kegan Paul, London.
GREEN, R. J. (1971) *Country Planning*, Manchester University Press, Manchester.
WELLER, J. (1967) *Modern Agriculture and Rural Planning*, Architectural Press, London.

Brief discussions of the geography of rural settlements and of rural planning are found in:

BAKER, A. R. H. (1969) The geography of rural settlements, in COOKE, R. U. and JOHNSON, J. H. (eds.), *Trends in Geography*, Pergamon, Oxford, pp. 123–32.

CLOUT, H. D. (1969) Planning studies in rural areas, in COOKE, R. U. and JOHNSON, J. H. (eds.), *Trends in Geography*, Pergamon, Oxford, pp. 222–32.

Changes in the land-use balance in Great Britain are considered in:

BEST, R. H. (1968a) Extent of urban growth and agricultural displacement in post-war Britain, *Urban Studies* **5,** 1–23.

BEST, R. H. (1968b) Competition for land between rural and urban uses, *Land Use and Resources: Studies in Applied Geography*, Institute of British Geographers Special Publication, **1,** 89–100.

BEST, R. H. and CHAMPION, A. G. (1970) Regional conversions of agricultural land to urban use in England and Wales, 1945–67, *Transactions Institute of British Geographers* **49,** 15–32.

COPPOCK, J. T. (1968) Changes in rural land use in Great Britain, *Land Use and Resources: Studies in Applied Geography*, Institute of British Geographers Special Publication, **1,** 111–25.

EDWARDS, A. and WIBBERLEY, G. P. (1971) An agricultural land budget for Britain, 1965–2000, *Wye College Studies in Rural Land Use*, **10.**

WIBBERLEY, G. P. (1959) *Agriculture and Urban Growth: a study of the competition for rural land*, Michael Joseph, London.

WIBBERLEY, G. P. (1961–2) Agriculture and land-use planning, *Town Planning Review* **32,** 77–94.

WIBBERLEY, G. P. (1967) The pressures on Britain's rural land, in ASHTON, J. and ROBERTS, S. J. (eds.), *Economic Change and Agriculture*, Oliver & Boyd, Edinburgh, pp. 155–67.

WIBBERLEY, G. P. (1972) Rural activities and rural settlements, mimeographed paper presented at the Town and Country Planning Association's Conference, London, 16–17 February 1972.

A wide-ranging discussion of increasing pressures on rural lands is found in:

BERESFORD, J. T. *et al.* (1967) *Land and People*, Leonard Hill, London.

Changing emphases in geographical investigation are outlined by:

WRIGLEY, E. A. (1965) Changes in the philosophy of geography, in CHORLEY, R. J. and HAGGETT P. (eds.), *Frontiers in Geographical Teaching*, Methuen, London, pp. 3–20.

CHAPTER 2

RURAL DEPOPULATION

CHANGES in the numbers of people living permanently in the countryside or making use of its resources on a temporary basis are central to rural geography and to the many problems of land management which are encountered in the countryside. Rural depopulation may be considered as the reduction of the absolute number of residents in a given area of countryside. Four main reasons make this statement not quite as straightforward as it may seem initially. First, the definition of phenomena that may be considered "rural" is a matter of great debate, depending, for example, on the range of variables analysed by each researcher. A sociologist would be likely to analyse different sets of data from those examined by a human geographer. Not surprisingly, verbal or quantitative conclusions reached by practitioners of different disciplines regarding "rural people" or "rural areas" will vary considerably. Second, the significance and precise meaning of the terms "rural" and "urban" differ in varying spatial, historic, and cultural contexts. Third, it is highly likely that modern definitions of "rural" conditions, in whatever context, would deviate from the administrative delimitation of "rural" areas prepared in the past for such official purposes as census-taking. Thus, after analysing employment characteristics, I. M. L. Robertson (1961) could conclude that only a quarter of the "rural districts" in England and Wales were still "rural" or "agricultural/rural" in the 1950's. The vast majority contained a predominance of urban and industrial features. A fourth problem involves variations in the definitions of "rural" and "urban" used by individual countries in their censuses or in other bodies of numerical data. International or cross-cultural comparisons derived from official sources labelled "urban" and "rural" must be treated with great caution.

For these and other reasons, "rural depopulation" is a nebulous phenomenon. Observers in the nineteenth century had no doubts that the distribution of population in western Europe was undergoing important changes during the period of industrialization. Urban growth undoubtedly formed the most dramatic expression of these changes, but in many country areas the resident population was decreasing over the same period. Just how rural contraction should be defined and measured was a matter of great debate. A. L. Bowley (1914), for example, criticized the official distinction between "rural districts" and urban areas in England and Wales. He insisted that only those "rural districts" which had registered population densities of three persons or less per 4 ha in the 1911 census should

8

really be considered as "rural". By applying such a definition of rurality, Bowley was able to show that population loss had occurred in the countryside between 1851 and 1881. He thus disproved the contrary suggestions put forward by W. Ogle (1889) who had restricted his analysis to population statistics at a county level from which figures for towns with more than 10,000 inhabitants were excluded. Many other researchers have devised quantitative expressions which, in their opinion, distinguish "rural" from "urban" phenomena in their own particular research context. Unfortunately no single definition has the merit of universal applicability in time and space.

TYPES OF DEPOPULATION

Absolute reductions in the number of residents in the countryside, however defined, can be produced by a variety of processes. The number of deaths in a settlement or rural area may exceed births over a given period of time. *Biological depopulation* will result if net immigration is insufficient to compensate for such losses (Pinchemel, 1957). Biological depopulation, in fact, is quite rare. It normally occurs only after protracted outmigration of young people has produced an unbalanced local age pyramid which is distorted by a predominance of old and middle-aged people. The reproductive potential of such a population is slight. Sufficient numbers of young people can rarely be attracted to migrate into such socially, and often economically, impoverished areas, and hence biological depopulation may occur.

D. Lowenthal and L. Comitas (1962) have shown that outmigration is undoubtedly the prime factor in rural depopulation. This view was confirmed in the English context by R. Lawton's (1968) analysis of population change at the registration district level between 1851 and 1911 (Fig. 2.1). Population decline was produced by an excess of net outmigration over natural gain in all rural areas where losses were registered. Only in central London was natural loss greater than net inmigration. However, analysis for individual settlements rather than larger spatial units might present very different results.

Various types of depopulation may be produced by net outmigration. P. Pinchemel (1957), in a detailed study of population change in part of northern France, drew the distinction between "occupational" and "non-occupation" outmigration. *Non-occupational outmigration* could be exemplified by a protracted movement of young people away from a densely populated agricultural area simply because of the extreme difficulty of finding any kind of employment in that area. Such movements would be general and not restricted to members of specific occupations. Non-occupational migration reflects a great pressure of population on local resources and suggests limited economic diversification in the particular area in question.

By contrast, *occupational outmigration* only affects members of specific rural groups. Farmers, or prospective farmers, might be forced to migrate in order to acquire a holding either if the demand to take over farms exceeded supply in their home area or if existing farms had to be enlarged (and hence reduced in number) if they were to continue to be viable. Landless labourers, dependent on wages for their

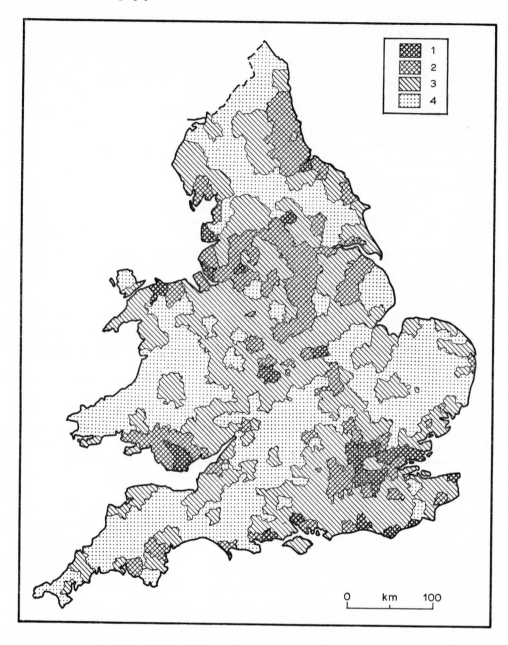

FIG. 2.1. Population change in England and Wales, 1851–1911. 1, Increase: net inmigration exceeds natural gain. 2, Natural gain exceeds net inmigration. 3, Natural gain exceeds net outmigration. 4, Decrease: net outmigration exceeds natural gain.

livelihood, might also suffer. They would be the victims both of agricultural depression, involving reduced profits and an inability to pay wages, and of technological change and mechanization which would diminish the needs for their services. Similarly, numerous craftsmen who served members of rural communities in the past would have to migrate in search of jobs elsewhere as the total population in their rural area declined. Their services might no longer be required, and their locally-made goods might be outpaced by cheaper, factory-made products.

Outmigration from the countryside will receive particular emphasis in the following discussion. Reductions in the numbers of farmers, farmworkers, and other rural employees may be identified readily from statistical sources to suggest a descriptive explanation of outmigration at a very high level of generalization. This would emphasize the labour-shedding characteristics of agriculture, rural manufacturing industry, and other activities when faced by technological change. It would assume fairly automatic human responses to the "repulsive" conditions of the countryside and the "attractive" ones of the town. Such a high level of generalization would suggest an inverse relationship between the volume of migrants and the distance they travelled between point of origin and urban destination.

An examination of case studies shows that such oversimplifications require considerable refinement. The following discussion will therefore be structured within the framework suggested by G. D. Mitchell (1950) who identified four main causes of rural depopulation. It will start with a broad consideration, in the context of England and Wales, of the reduced demand for agricultural labour and the changing economic structure of rural settlements, with special regard to manufacturing industries and services. It will then focus down from such generalizations to review social and other deficiencies at the local or community level that may encourage outmigration. Attention will then be drawn to real and apparent reductions in the economic and non-economic attractions of rural life, and to highly subjective issues involving the rural migrant as a decision-maker.

REDUCTIONS IN AGRICULTURAL LABOUR

Reduced demands for agricultural labour have been viewed as forming an essential component in the complex, interlinked processes leading to rural depopulation in England and Wales. Changes in the agricultural sector may be identified quite readily from occupational statistics. However, reductions in the number of primary producers should not be overemphasized to the exclusion of other groups that have also declined.

Enclosure, land reclamation, and agricultural improvement retained large numbers of labourers on the land in the eighteenth and early nineteenth centuries before the mechanization of agricultural work. J. D. Chambers (1953) noted that whilst early enclosure activities provoked depopulation (for example, linked to the conversion of arable farming to sheep grazing in the sixteenth century) this was not the case in the nineteenth century when another form of enclosure replaced open arable strips with much larger enclosed fields. Studies of early nineteenth-century census material suggest no direct correspondence between the occurrence of enclosure and the outmigration of rural population. In fact, as stretches of uncultivated land were brought into use and improved the local farm population increased. New agricultural practices, associated with the new husbandry, called for more labour—not less—to work in fields, barns, stockyards, and dairies. Such crucial activities as hedging and ditching to maintain newly installed field boundaries around enclosed property demanded larger inputs of agricultural labour than had been required in the past. Contrary to popular opinion in the late nineteenth century, enclosure and land improvement generally helped to retain people in the land rather than driving them away to the towns.

The absolute number of people living in rural districts in England and Wales continued to rise until 1851 and thereafter has remained fairly stable in absolute terms (Fig. 2.2). However, the total population of the country has

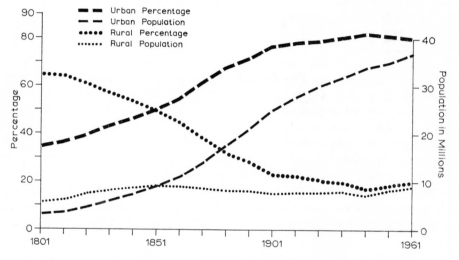

Fig. 2.2. Changes in the absolute numbers and relative proportions of urban and rural populations in England and Wales, 1801–1961.

increased dramatically so that the relative importance of residents in officially designated "rural districts" fell from 50 per cent in 1851 to only 20 per cent in the 1960's. It must, of course, be remembered that many "rural districts" display urban characteristics.

No individual county exhibited a decline in its total population until the mid-nineteenth century when decreases were noted in Wiltshire and Montgomeryshire. However, at a finer scale of analysis, absolute declines had been registered from 1821 onwards in an ever-increasing number of rural parishes and communities. Such local decreases were often not detected by casual observers of county totals, who wrote of the "alleged depopulation" of the countryside. In the 100 years after 1851, population losses were recorded consistently for each intercensal period in Cumberland and counties in central and western Wales. Uninterrupted decline was not experienced in other counties, but rather complicated intermixtures of gains and losses which varied considerably between different intercensal periods.

The agricultural labour force of England and Wales increased in absolute numbers during the first half of the nineteenth century, even though 1815–50 was a period of depression. J. Saville (1957) remarked that: ". . . . despite gloom, it is probable that both agricultural employment and production continued to increase although the demand for labour from agriculture was insufficient to absorb a rural population that was growing rapidly" (p. 9). In 1861 a peak of 1.9 million farmworkers was reached. Thereafter the number of regular wholetime workers fell to 990,000 in 1901 and a mere 295,000 in 1966. The total would

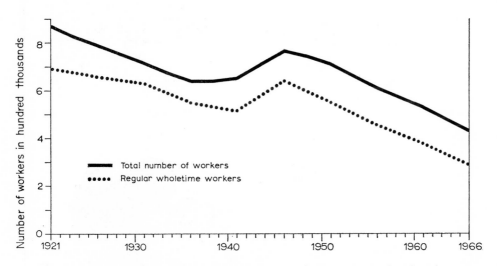

FIG. 2.3. Changes in the number of agricultural workers in England and Wales, 1921–66.

be greater if seasonal workers were also included. Numbers had, in fact, risen during World War II (Fig. 2.3) when far greater emphasis was placed on the domestic production of foodstuffs. Figure 2.4 conveys something of the dramatic decline in the density of farmworkers per 400 ha of agricultural land which occurred between 1921 and 1966 in Great Britain.

The marked reduction in the number of workers on the land in England and Wales was associated with important changes in the structure of the hired labour force. In the early nineteenth century seasonal demands for farm labour had been met by gangs of men, women, and children which were organized by private gang masters until the system was controlled by he Gangs' Act of 1867. The structure and size of the agricultural labour force was further modified by the

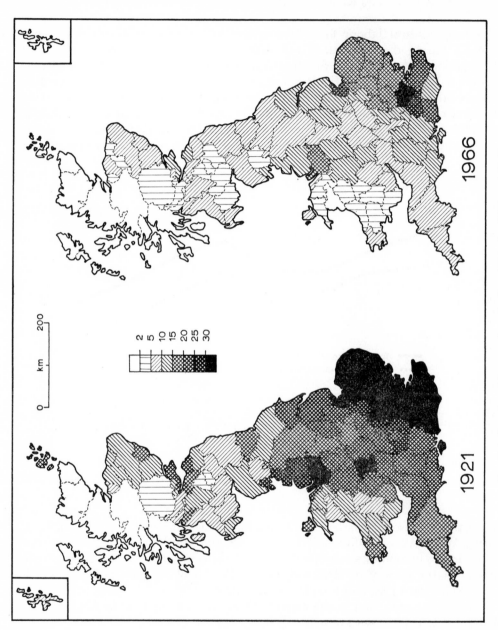

Fig. 2.4. Density of farmworkers in Great Britain, 1921 and 1966.

Education Act of 1870 which provided compulsory primary education and thereby removed children from the workforce. Between 1861 and 1901 the numbers of young people under the age of 20 working on the land fell from 428,000 to 195,000 and the number of women from 436,000 to a mere 52,000.

Many late-nineteenth-century commentators attributed reductions in the farm labour force to English free-trade policy and especially to the free entry of corn. G. B. Longstaff (1893) declared that: "free trade has almost certainly fostered the movement from country into town in two ways: its forces have been at once attractive and repulsive. By lowering the prices of agricultural produce, more especially wheat, it has made farming less profitable, turned cornfields into pastures, and so *driven* the people from the country. By increasing the volume of foreign trade it has increased the demand of the manufacturing districts for labour and has so *drawn* the people into the towns" (p. 388). This line of argument was highly generalized, emphasizing "push" and "pull" forces and placing stress on economic factors.

However, Longstaff's approach changed later in his report. He found that depopulation was encountered both in developed countries "from the bogs of Galway to the waving cornfields of Hungary" (p. 399) and in newly-settled parts of the world, such as Canada, the United States, and Australia. Contracting agricultural employment could not be attributed to specific types of government, land tenure, or fiscal policy. Longstaff stated: "I believe the main causes to be two, of which one may be called sentimental and the other economic" (p. 399). The sentimental, or social, cause was the "lack of pleasure" for the country dweller who dreamed "to get away from the country". Longstaff continued: "I believe this, which I have called the sentimental cause, to lie at the root of the matter, but it is, all said and done, of the nature of what medical men term predisposing causes; we have now to consider the exciting cause" (p. 413). This was summed up by improvements in communications. Rapid locomotion and an inexpensive post-office and telegraph system made possible a cheap press. Primary education and the spread of literacy meant that newspaper information could be absorbed so that "men learn where there is a demand for labour and are directed to it" (p. 413).

Lord Eversley (1907) emphasized both the "pull" effects of the town as well as the "push" of the countryside, and also the importance of social as well as economic phenomena. "The explanation of the great exodus of labourers from rural districts during the 20 years 1861–1881 was undoubtedly to be found in the great prosperity and the general rise of wages in the manufacturing and mining districts: in the fact that labourers were tempted by the higher wages in the towns, and also work on the railways, and in the rural police, to give up farm work. It was also in part due to a growing disinclination to farm work among labourers in rural districts, to the absence of opportunities to them of rising in their vocation, and to a desire for the greater independence and freedom of life in towns" (p. 280).

G. Ahlberg (1956) has shown that the labour-shedding function of agriculture is more complicated than previously believed. He argued: "were it simply a question of the release of excessive manpower, the migration losses ought to be much the same from year to year but in fact they have been subject to significant variations which have an obvious connection with business cycles" (p. 10). Thus only 15,000 workers were released from Swedish farming each year in the depressed years of the early 1930's, but in the "boom" at the end of the decade the annual figure rose to over 40,000 and over 60,000 in the post-war recovery years in the late 1940's.

Crisis conditions in agricultural economies also provoked major outflows of farmworkers in search of alternative employment. Potato disease and subsequent famine during the 1840's affected not only Ireland but many parts of upland Europe, and thereby intensified the volume of outmigration. In a similar fashion, devastating attacks of disease after 1865 shattered previously profitable agricultural economies built on vine production and silk-worm rearing and accelerated rural/urban migration in parts of southern France.

Another aspect of agricultural change of relevance to rural depopulation involves reductions in the mean size of farming household. In the past extended families with several generations living under one roof were encountered in many agricultural environments in Britain and in Continental Europe. In advanced agricultural areas, such as east Yorkshire, large farming households were found for another reason. This involved considerable numbers of young male farm servants "living in" households with as many as fifteen persons in the mid nineteenth century. The age/sex structure of such households was severely distorted. Both forms of extended farming household have declined in importance in the present century. June Sheppard (1961) showed that many rural households in east Yorkshire declined to one-half or one-third of their previous size over the past 100 years or so since apprentices, farm servants, and maids no longer "lived in". The total number of farming households in an area may remain constant but the total population involved may decrease considerably.

The progressive release of labour from agriculture represents only one form of occupational outmigration. Rural craftsmen producing goods and providing services for country dwellers were also affected as the economic structure of rural settlements changed in the past couple of hundred years.

THE CHANGING ECONOMIC STRUCTURE OF RURAL SETTLEMENTS

The distribution of manufacturing activities in the now developed countries prior to the industrial revolution was linked to sources of water power, charcoal fuel, local supplies of metal ores, and other raw materials. Textile industries flourished in many rural areas. Similarly, the charcoal-based iron industry prospered where ore deposits and timber supplies were found in close proximity. For example, forges functioned in numerous upland parts of France on the eve of the Revolution (Fig. 2.5). Regional specialisms in manufacturing activity were found throughout the countryside. Augustus Petermann's maps of industrial activities in the British Isles were prepared to accompany the 1851 census and reveal patterns of industrial location that were very different from those of today. Figure 2.6 simply shows the distribution of a variety of textile producing activities. An

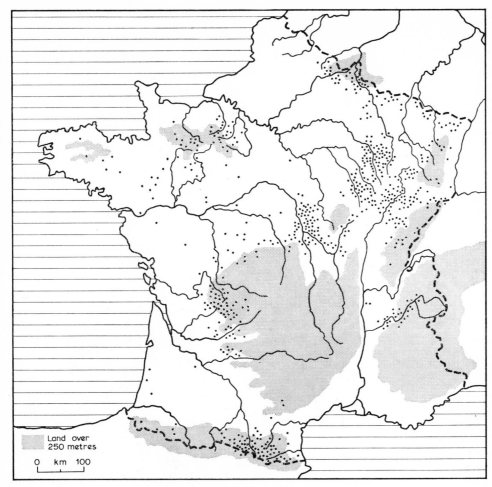

FIG. 2.5. Distribution of forges in France, 1789.

extremely complex picture would emerge if all rural industries depicted by Petermann could be collated on a single, small-scale map.

The decline of rural manufacturing both contributed to and resulted from the
more general contraction of population in the countryside. As the number of
agricultural workers decreased so the demand fell for the services of tailors, blacksmiths and other craftsmen, and also for locally-produced goods such as textiles,
furniture, and metalware. In any case, such products had to compete with factory
goods distributed along turnpike roads, canals, and railways. Manufacturing and
service enterprises had formerly guaranteed a broad-based social structure in
many rural communities. Such activities either disappeared completely, as was
the case of thatchers and saddle-makers, or were adapted to meet new needs.
Thus boot-makers became shoe-repairers, and many blacksmiths and wheelwrights
became garagemen.

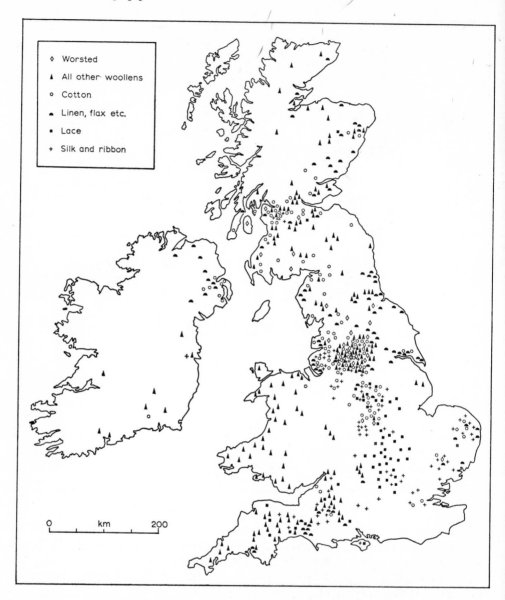

FIG. 2.6. Distribution of the textile industry in the British Isles, 1851.

Spatial changes frequently resulted as industries that had formerly depended on sources of energy (water, charcoal) and raw material (home-grown corn, local wool, or leather) became increasingly reliant on new forms of energy provided by the coalfields and on raw materials imported from abroad. Industries which had once been distributed widely in the countryside occupied new locations at ports and major concentrations of urban population. Corn-grinding, for example, ceased to be a truly rural activity in England during the final quarter of the nineteenth century as large mills were opened in county towns to deal with home-grown supplies and in the ports to handle grain imported from overseas.

John Saville (1957) provided other illustrations. The furniture industry of the Chiltern Hills had involved 160,000 chair-makers scattered in villages in the 1880's but by the end of the century it had become concentrated in factories in the town of High Wycombe. Similarly, the paper-makers of Devon abandoned their remote workshops alongside fast-flowing streams and concentrated their activities at far fewer sites. Technological innovation was intimately associated with urbanization, and many old-established craft industries in rural areas simply disappeared as the construction of the railway network destroyed the historic isolation of country areas and permitted urban-produced goods to be distributed easily.

Not only had the agricultural population of England and Wales contracted by 1900, but numbers of rural artisans and service workers had fallen as well. June Sheppard's (1962) detailed analysis of population changes in three villages in east Yorkshire provides a hand specimen of this kind of change. The countryside continued to support farmers and agricultural workers, admittedly in much smaller numbers than before, but very few representatives of industry, service trades, and other sectors of the rural economy remained. A process of "ruralization" had set in.

R. J. S. Hookway (1958) emphasized that the secondary and tertiary population still resident in the English countryside formed a likely group for future outmigration. He believed that "a proportion of the secondary population of the small villages and even the smaller market centres will be displaced if our standard of living goes up. Already their productivity . . . is small and as well as unemployment there is a great amount of underemployment, often associated with a multiplicity of part-time occupations. It is amongst the secondary rural population that one now finds the poorer elements of the rural society, and the trading and servicing tentacles stretching out from the large market centres are an ever increasing threat to their livelihood" (p. 212).

It could be argued that the distinction between "town" and "country" became more apparent during the late nineteenth and early twentieth centuries than ever before as large numbers of country dwellers migrated townwards but very few town dwellers returned to live in the country. This counter-flow was not to be produced until public transportation was improved in the inter-war years and suburbia overflowed into the countryside. Rural/urban contacts were intensified after World War II as increased leisure time, more affluence, and higher rates of car ownership permitted city folk to make greater use of the countryside for residence, recreation, and retirement.

POOR RURAL FACILITIES

So far, attention has been drawn to the role of labour-shedding activities in contributing to outmigration from the countryside. But in addition to such general matters one must consider deficiencies in the specific localities inhabited by rural dwellers. There were persistent reports throughout the nineteenth century of shortages of cottages for housing agricultural labourers in rural Britain. Housing conditions were generally inadequate in spite of rebuilding on some country estates. As already explained, several generations in a farming family frequently had to share a single house. Even now rural areas still lag behind the cities in receiving piped water, electricity, and mains-drainage facilities.

It is certainly correct to stress the role of such deficiencies in stimulating out-migration from the countryside, but it is clear that precise correlations should not be expected between rates of depopulation and standards of service provision. In a detailed study of conditions in Somerset, H. E. Bracey (1958) noted that "some enthusiasts would have us believe that rural depopulation can be explained wholly in terms of the lack of bathrooms and inadequate village school facilities" (p. 67). Population loss in rural Somerset between 1931 and 1951 was found in the eastern lowland, arable land, the central marshy levels, and the western areas of Exmoor and the Brendon Hills. In the west the physical environment is harsh and agricultural living conditions are hard. This is not the case, however, in the central and eastern parts of the county. Bracey proceeded to examine public utilities (piped water, electricity, gas, sewerage, and refuse collection) in order to see if correlations could be established with rates of population change. In a general sense depopulating areas were less well provided with facilities than were expanding villages. But very important local anomalies were recorded so that one in every ten parishes with increasing population was found to be very poorly served. Bracey concluded that other social criteria needed to be investigated to help explain the observed patterns of change. These would include "those elusive qualities, initiative and leadership, which may be found in the smallest parish and are very frequently present where there is a significant number of newcomers" (p. 73).

A similar conclusion had been reached by G. D. Mitchell (1950) in his study of rural parishes in south-west England. With reference to poor housing conditions and the absence of electricity and piped water, he noted that: "in some parishes there is a universal and insistent demand for them, whilst in others they appear to arouse little interest; this is largely a function of the different social outlooks peculiar to different localities" (p. 5). Mitchell stressed that settlements in the area he investigated possessed markedly individual characteristics. He did not consider it possible to regard the area as a single, or even as part of a single, rural society except in a rather superficial sense. Instead, several societies could be recognized. These might be coterminous with village or parish boundaries or they might overlap to cover several parishes or parts of several parishes.

Four "fairly distinct types of rural ethos" (p. 12) could be discerned. At one level a distinction could be drawn between "socially integrated" societies and those "undergoing a process of disintegration", in other words between localities with discernible harmonies or disharmonies in institutional life. Cutting across these two types were "closed" and "open" societies. In "closed" societies newcomers would be regarded as "foreigners", but in "open" societies they would be easily assimilated into local activities.

The "open integrated society" was considered by Mitchell to be the most desirable. It would be characterized by a fairly large village with a stable or growing population, a distinct sense of civic pride, an active parish council, and a range of local entertainment and educational facilities. Although not quoting figures for "threshold populations", Mitchell was describing what would now be referred to as a "key village" that would be sufficiently large and well equipped to provide services for the resident population in surrounding smaller settlements. Open integrated societies were not numerous in the study area but closed integrated societies were very common. No overt disharmony was in evidence but these villages were very traditional in their outlook. Some were dominated by introspective religious groups. Open disintegrating societies were characterized by innovation and population growth, e.g. following the opening of a new factory. But social organizations were unstable and a considerable turnover of population took place. Closed disintegrating societies were characterized by marked depopulation and an absence of local leadership.

Mitchell considered that a sense of "community" was essential if rural residents were to remain living in the countryside. "If this lack of integration is not remedied there is nothing but meagre economic incentives to keep people in the rural area" (p. 84). This important sociological study placed valuable stress on spatial differences in local conditions which geographers, all too often, tend to overlook. Whilst the establishment of more and better housing, social amenities, and rural utilities might be accomplished, Mitchell found it difficult to see how these alone would re-establish a sense of community feeling. He concluded that "fewer, but larger villages seems to be the solution to many of the social problems" (p. 84). This point will be taken up again at a later point in the discussion.

REAL AND APPARENT REDUCTIONS IN THE ATTRACTIONS OF RURAL LIFE

Underlying the various causes of outmigration discussed so far is the basic assumption that rural residents acquired knowledge of conditions other than those in their home community and were then able to make use of such information to compare their own living conditions with those reported from the city. During the present century, mass media of communication, such as newspapers, radio, and television, have diffused this type of information. Rural dwellers have become aware that urban jobs are generally better paid than work in the countryside and also that living conditions in towns are normally better than in rural settlements. H. E. Bracey (1970) emphasized seven attractions of urban areas which might encourage rural people to migrate citywards. These were: better shopping facilities; a greater range of social amenities; better housing; improved welfare; better transport; better education facilities; and a wider range of social and intellectual contacts (p. 34).

Such advantages were given quantitative expression in the results of surveys in north-eastern England (House, 1965). These showed that 72 per cent of the folk who had migrated from the countryside stated "employment" as the prime reason for their move; 25 per cent moved when they became married; and 3 per cent moved for reasons of education. A more detailed analysis of reasons quoted by folk intending to migrate from the countryside (Table 3) again placed employment in the leading position (36 per cent). When social reasons were summated

R.G.—B

(shopping, social amenities, transport, and education) they represented 49 per cent of all replies. Only 3 per cent of respondents stated the search for better housing as the main reason for rural/urban migration. D. Hannan's (1969) analysis of the migration intentions of Irish rural youth emphasized the importance of restricted rural employment as a factor provoking outmigration. But such conditions were perceived and interpreted in varying ways by different individuals. "*Beliefs* about one's ability to fulfil 'economic' type *aspirations* locally" were the dominant variables in the migration plans of young people (p. 202).

TABLE 3. REASONS QUOTED BY RURAL DWELLERS IN NORTH-EAST
ENGLAND CONTEMPLATING MIGRATION TO URBAN AREAS

	(%)		(%)
Employment	36	Better climate	3
Better shopping	15	Better housing	3
Better social amenities	14	To retire	2
Better transport	10	Promotion	1
Better education	10	Health	1
To be near relatives	3		

Source: HOUSE, J. W. (1965), *Rural North-east England: 1951–61*, Department of Geography, University of Newcastle-upon-Tyne, Papers on Migration and Mobility **1**, p. 45.

A detailed survey of 250 individuals who left agricultural employment in Gloucestershire after 1950 emphasized the perceived disadvantages of farmwork and showed how labourers in varying age categories stressed differing disadvantages (Cowie and Giles, 1957). Poor pay was quoted most frequently, being mentioned as the leading reason for leaving the land in 30 per cent of all replies. Long and uncertain hours were stressed in 10 per cent. Other frequently mentioned reasons included failing health, redundancy, poor working conditions and accommodation, a lack of prospects for advancement, and the problem of the tied cottage. This last feature was seen as a weapon or enticement in the hands of the employer, and imparted a sense of insecurity in the farmworker. Low pay and long hours were listed in leading positions by 40 per cent of all the men interviewed, but represented 60 per cent of answers quoted by men between 16 and 25. This is the marrying age when many farmworkers may seriously consider their "position" for the first time. These two reasons took on steadily decreasing importance as age increased, with problems of health becoming of greater significance, together with redundancy and complaints over the tied-cottage system. Amongst men over 55 years of age the tied-cottage issue took on the dimensions of "an element of fear and insecurity" (p. 87).

The results of surveys undertaken by French sociologists on how rural dwellers interpret the concept of "the town", view the rural/urban migration issue in a rather different way. Interviewees were asked for snap answers but were then given more time to think out their replies. The snap answers (Table 4) showed that most countryfolk interpreted the town as a place offering an attractive source

TABLE 4. SPONTANEOUS IMPRESSIONS OF "THE TOWN"

	All responses (%)*	Men (%)	Women (%)
The town seen as a place of:			
work	37.7	48.0	20.0
entertainment	22.5	16.0	33.3
noise and movement	15.0	12.0	20.0
freedom	10.0	8.0	13.3
social constraints	10.0	12.0	6.7
no opinion	22.5	20.0	26.7

* Total exceeds 100 per cent since some answers were dual responses.

Source: RAMBAUD, P. (1969), *Société rurale et urbanisation*, Seuil, Paris, p. 21.

of employment. This view was particularly strongly represented amongst replies from men. Women emphasized the entertainment facilities which the town offered, its noise and bustle, and added freedom. Each of these features contrasted with the boredom, quietness, constraints, and routine of working on the land. The thought-out answers were rather different (Table 5). Educational advancement came at the top of the list and was emphasized more strongly by men than by women. The chance of obtaining more leisure came in second position and was mentioned more by women than by men. These replies provide valuable evidence on the perceived restrictions of rural life, in terms of employment and social advancement, and on the anticipated advantages to be gained from urban living.

TABLE 5. THOUGHT-OUT IMPRESSIONS OF "THE TOWN" (PROPORTION OF ALL RESPONDENTS MENTIONING SPECIFIC TOPICS)

	All responses (%)	Men (%)	Women (%)
The town seen as a place:			
of education	72.5	76.0	66.7
allowing free time	47.5	44.0	53.3
for earning money	40.0	40.0	40.0
for social advance	37.7	36.0	40.0
permitting new friendships	37.7	36.0	40.0
of culture	37.7	24.0	60.0
of noise and lack of air	20.0	20.0	20.0
equals the factory	17.5	24.0	6.7
of anonymity	15.0	12.0	20.0
to counteract boredom	12.5	20.0	0.0

Source: RAMBAUD, P. (1969), *Société rurale et urbanisation*, Seuil, Paris, p. 22.

G. D. Mitchell (1950) laid particular stress on the social advantages to be gained from living in the city. He believed that "the economic attractions of urban employment play a secondary role to the social attractions which are a function of the continued penetration of urban ideas and outlook presented by the press and radio" (p. 4). But urban ideas had been diffused in a variety of other ways in the past. Traditional rural economies in many areas involved forms of activity to supplement farming so that a large rural population might be supported from limited agricultural resources. In some instances craft industries or outworking would flourish during the winter months when agricultural work was slack. Spinning, weaving, nail-making, and other crafts operated in the French Massif Central, for example, as peasant families received raw materials from urban entrepreneurs in surrounding cities. Links between rural and urban areas were forged, and city ideas were diffused.

Sometimes it would be necessary for rural labourers to move elsewhere to obtain seasonal work. This might involve purely local movements as labourers sold their services to farmers in nearby regions for harvesting cereals, hay, or vines. In other instances they moved to urban areas as tinkers, sweeps or tradesmen (Fig. 2.7). The movement of men from the Massif Central to work as builders' labourers in Paris, Lyons, and other French cities during the nineteenth century provides a good example of temporary migration which later hardened into definitive outmigration. Such temporary movements were very widespread in Europe. They involved a transmission of ideas into the remote countryside from cities or from more advanced farming regions.

D. Pinkney (1953) studied the diffusion of information linked to movements of Limousinants to Paris. Because of "long experience of seasonal migration, the thought of going to the city was neither strange nor frightening" (p. 5). Living conditions in the Massif Central were harsh and wages were low. They certainly acted as repellent features, but migrants also needed to be attracted to their specific destinations. Parisian newspapers carried advertisements for builders' labourers, but they were not circulated in the Massif. Personal letters were far more important. Late in the nineteenth century, cheap railway excursions to the capital provided another means of diffusing urban information. But the most important medium was the temporary migrant himself when he returned to his home village after a season in Paris. Contemporary observers noted how he formed the centre of interest in his home village. Neighbours surrounded him, anxious to learn how much he had earned and to hear about what he had seen in the capital. Even more eloquent than his embellished descriptions of the city he had helped to build were the francs that he brought home. If a migrant had a successful season, the village would soon know of it, and his earnings would be common knowledge for all to marvel at and for boys and young men to aspire to equal by taking the road to Paris. Traditions of migrating from the Massif Central to take up jobs in Paris still occur, and have been followed by family members for generations.

This type of human "migration system" has been identified in many other parts of the world. Thus G. A. Hillery and J. S. Brown (1965) noted that "the southern Appalachians is not a region in the sense of its parts belonging to the same migra-

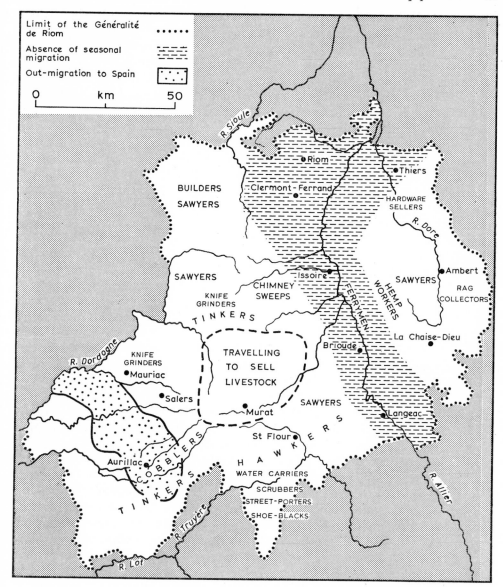

FIG. 2.7. Types of occupation practised by temporary migrants from the Massif Central during the eighteenth century.

tion systems. Rather it is a collection of fringes, or, as it has often been put, of 'backyards'. The various parts of the southern Appalachians are closely tied to non-Appalachian areas in one of the most crucial ways possible; their people migrate to these areas" (p. 47). Members of a given extended family would migrate to the same place and then stay with their kin who provided "havens of safety" or "a socio-psychological 'cushion' for the migrant during the transitional phase"

(Brown *et al.*, 1963, p. 48). As a result, streams of migration were "running in well-worn riverbeds" (p. 52). Intervening rural areas and small towns were avoided, as was the case with migrants from the Massif Central, and the rural migrants headed straight for distant cities. "The kinship structure provides a highly persuasive line of communication between kinsfolk in the home and the new communities which channels information about available job opportunities and living standards directly, and most meaningfully to . . . [rural] families. Thus, kinship linkage tends to direct migrants to those areas where their groups are already established" (Brown *et al.*, 1963, pp. 53–54).

Examples of this kind of "friends and family" effect in diffusing information may be recognized in other contexts. D. Hannan (1969) showed that "although industrialization and urbanization has been very slow to develop in Ireland, most rural families . . . are highly integrated into urban industrial relationship networks through their contacts with immediate family members, other relatives, friends and neighbours living and working in industrial areas in Britain, in the US, and in Ireland" (p. 195). Similarly, E. Kant's (1951) analysis of the migration catchment of Budapest in the early twentieth century (Fig. 2.8) revealed a different pattern from the simple concentric, distance-decay organization one might have expected. This distortion was due to a cultural "migration system" which attracted large numbers of migrants from the Hungarian exclave in Transylvania east of Budapest.

In addition to information diffused through kinship links, comparisons between rural and urban conditions would be facilitated by various other social processes. The spread of primary education in country areas in the past 100 years served to teach country children of an alternative urban environment and way of life. John Walter put the point succinctly in 1889 when he remarked that, as a result of education, "boys got to think that broadcloth was better than fustian and girls thought a little finery suited them best" (Longstaff, p. 415).

D. Hannan's (1969) study of migration differences in rural Ireland confirmed that "migration is highly selective by education level achieved. Only 12 per cent of the secondary educated intended to remain in the home community, compared to 25 per cent of the vocationally educated and 38 per cent of the primary educated. The level of education is also closely related to the level of occupational and income aspiration of respondents, and, consequently to the level of frustration of these aspirations" (p. 207). Only 33 per cent of the secondary educated believed that they could fulfil their occupational aspirations locally, but this was true of 48 per cent of the vocational- and 65 per cent of the primary-educated. Criticisms of local amenities and social provisions became increasingly numerous as the level of education increased.

The same investigation provided ample supporting evidence for the seventh "law of migration" that had been enunciated by E. G. Ravenstein in 1885, namely, that "females are more migratory than males". Hannan emphasized two main reasons for this condition. First, farming is a sex-selective occupation. Male heirs normally inherit farms and see their future as intimately bound up with farmwork. The situation is just the reverse for farm girls in the developed world who look to the city for their future. (In other parts of the world, however,

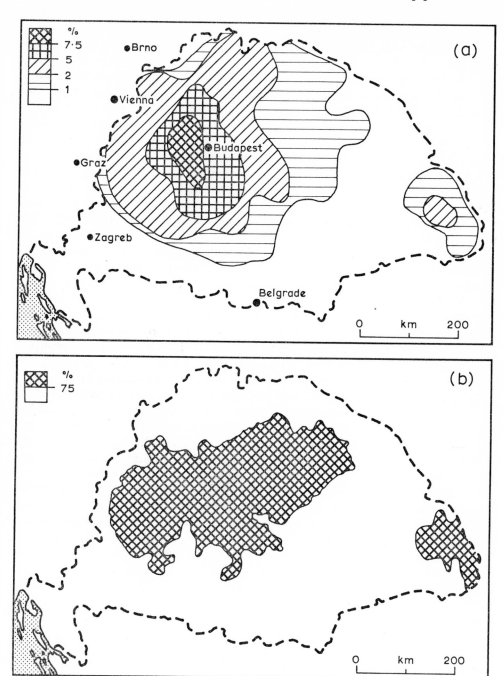

FIG. 2.8. (a) Migration field of Budapest, 1911. (b) Magyar-speakers as a proportion of the total population.

traditional cultural attitudes may keep women within the rural family, and hence more men would migrate than women.) Second, in many developed countries there is a clear cultural bias in favour of educating country girls to a higher level than boys. This is an attempt to overcome the difficulties experienced by rural girls in obtaining local jobs of a suitable status level. Education enables them to tap better occupational and marriage opportunities elsewhere. Marked differences in status and occupational aspirations exist between boys and girls who have had more than primary education. D. Hannan noted that only 10 per cent of the vacationally educated girls in his sample intended to stay in their home settlement but more than three times that proportion of equally educated boys intended to do so.

J. W. House (1965) found that female outmigration from the countryside tended to take place at certain key ages, such as immediately after leaving school and in the early twenties when the search for an urban marriage partner forms a strong motive for outmigration. Women migrants also tend to move at an earlier age than men. Wendel's (1953) study in Sweden showed that the mean age for the first departure of men from the countryside was 25 years, but for women it was 22 years. Rural women also tended to migrate greater distances than men to obtain both tertiary employment, that was not available in the country, and a city-dwelling husband.

Compulsory military service, operating in various ways over the last 100 years or so, provided a mechanism for introducing country boys to an urban way of life, this time by taking them away to barracks which were often in peri-urban locations.

Each new phase in transport technology facilitated rural outmigration in three main ways. First, rural labour was required for highways and railways to be constructed. Gangs of navvies moved from their home villages as they built the routeways. Second, the new forms of communication—and especially the railways—provided easier means of moving to the city. Third, they allowed a diffusion of urban ideas into the countryside as travellers brought back information and as newspapers were sent cheaply and rapidly. John Saville (1957) has rightly stated that the importance of the railway in Britain was in accelerating migration from the land—not in initiating it. Flows of population to English towns were strongly established well before the opening of the Manchester–Liverpool railway in 1830 ushered in the "railway age". Increased mobility certainly resulted, but it might be argued that the real significance of the railways stemmed from their creation of a nationwide market for goods, news, and ideas rather than from their ability to transport rural outmigrants.

PERCEPTION AND DECISION-MAKING IN DEPOPULATION

In recent years the various problems linked to trying to explain human migratory movements have become particularly acute. Planners, demographers, geographers, and others have been concerned with trying to predict migration so that housing, education, and other facilities may be provided at appropriate scales both in depopulating rural areas and in growing urban ones. Experimental work during the 1950's for predicting volumes and patterns of migration linked largely to a consideration of likely migration distances, met with little success. J. Wolpert (1965) noted that "it must be admitted that the gravity model and its elaborations appear to lose explanatory power with each successive census. When population flows are disaggregated, the need becomes greater to determine unique weightings for areas and unique distance functions for sub-groups of in- and out-migrants" (p. 159). In short, models which emphasize the distance factor between point of origin and eventual destination in human migration are inadequate. New approaches are required which consider the perception and the decision-making processes of individual migrants or sub-groups of migrants.

In the past, attempts to explain outmigration operated at a relatively high level of generalization, emphasizing, first, the significance of linear distance between point of origin and destination, and, second, assumed qualitative differences in places whereby the countryside would be considered by the potential migrant as "repulsive" and the city "attractive". Such assumptions have been accepted, but little research has been undertaken regarding the behaviour of migrants and would-be migrants as they react to a wide range of influences. These include both proximate or personal causes (an individual's aspirations, his attachment to or alienation from his home community) as well as impersonal social structural forces (such as differences in degree of education, the sex of the candidate, and the geographical remoteness of the place where he lived). In many instances, parental attitudes for encouraging, or actively discouraging, migration by their children would be important, as would the social obligations of the individuals involved. For example, one child in every family might be expected to stay on the farm and look after his parents in their old age. But in D. Hannan's (1969) study the desire of young rural-born Irish people to satisfy their economic aspirations took precedence over family obligations. Some respondents were so alienated from their home community that they would migrate irrespective of their beliefs about local opportunities.

J. Wolpert (1965) has argued that three key concepts need to be considered in discussing migrant behaviour. The first is "place utility" whereby each decision-maker, or would-be migrant, ranks the alternative destinations to which he might move in terms of the increased satisfaction he expects to achieve through moving. Human beings are "intendedly" or "boundedly" rational individuals who try to rationalize decisions according to the information at their disposal. But they have a very limited ability to use information, to perceive, calculate, and predict. They also have a very imperfect knowledge and understanding of the environment in which they live and the various alternative environments to which they might move. Nevertheless, decisions are taken, individuals make choices, and differentiate between alternative courses of action (in this case whether to leave the countryside or to stay, when to depart, and to which urban destination to go).

Each decision-maker has his subjective concept of "place utility" based on his own individual aspiration level which is adjusted on the basis of personal experience and the flow of information from family members, friends, and the mass media. The individual sets up a threshold of net utility which allows him to distinguish between success and failure. So far a twofold process of cause and effect leading to rural outmigration has been suggested involving, first, the perception of sets of stimuli (including the varying merits of town and country living), and, second, reacting to the perception of these stimuli. Such a reaction may involve immediate outmigration, but equally well may involve postponement until some future time, or, alternatively, a change in aspiration and readaptation to the present place of residence.

The second major theme meriting consideration involves the "search" for information. In an historic rural context, information would be almost totally derived from local sources with, perhaps, just occasional inputs of information from a national level in the form of royal decrees. Gradually, the development of new forms of transportation and communication (highways, railways, cheap post and telegraph, newspapers, radio, and television) facilitated the movement of individuals as purveyors of information and also the flow of news by various media which did not require face-to-face contact. Each individual in the modern world samples from the vast range of information which is theoretically at his disposal. Such a sampling process depends on a very wide range of social parameters, including the individual's age, education, family background, and aspiration. Only a limited section of the mass of information is selected by the individual as relevant to his "action space", namely that section of the universe of space and time in which he conceives that he might comfortably exist. "Action space" is a highly personal concept and is as much a product of each decision-maker as of the environment in which he lives. In this sense, "environment" would not be restricted to the settlement inhabited by the decision-maker since mass communications, contact with friends and relatives, variations in education and travel, and a host of other processes provide individuals with varying ranges of information from local, national, and international sources from which to sample.

A third major concept involves the "life-cycle approach" comprising important human factors such as education, marriage, the search for a niche in life, promotion in employment, retirement, and a host of other features which affect any decision regarding outmigration. In the case of unattached migrants a factor such as the quest for promotion in employment may be considered in a very straightforward way. But once the migration of family groups is involved it is necessary to view not only the merits (or otherwise) of moving for the head of household but also the likely repercussions on the wife and children. Human migrations have certain characteristics in common with other aspects of flow in that they reflect connectivity between places, but they are different from all other flows for two main reasons. First, the object which is transported is itself active and decides when it, and its family, should move and where they should go. The agent generates its own flow. Second, human migration involves upheaval and social disengagement of varying intensity according to the size of the household and also according to the closeness of the links of its social network.

The preceding discussion has demonstrated something of the complexity of rural–urban migration of which aspects have been investigated at a variety of scales by practitioners of each of the social sciences from psychology, through sociology and economics, to human geography. A strong case could be made for pluri-disciplinary investigations of such movements. But it might be suggested that these should perhaps be conducted in the urbanizing, developing parts of the world rather than in Europe or North America where reverse flows of population are now extremely important as urbanites rediscover the countryside and migrate to it for residence, recreation, and retirement, thereby provoking important changes in its social composition.

REFERENCES AND FURTHER READING

A comprehensive study of rural depopulation is provided by:

SAVILLE, J. (1957) *Rural Depopulation in England and Wales, 1851–1951*, Routledge & Kegan Paul, London.

A general consideration of outmigration is presented by:

LOWENTHAL, D. and COMITAS, L. (1962) Emigration and depopulation: some neglected aspects of population geography, *Geographical Review* **52**, 195–210.

Historical aspects of rural depopulation are discussed in:

CHAMBERS, J. D. (1953) Enclosure and labour supply in the industrial revolution, *Economic History Review* **5**, 319–43.

LAWTON, R. (1967) Rural depopulation in nineteenth century England, in STEEL, R. and LAWTON, R. (eds.), *Liverpool Essays in Geography*, Longmans, London, pp. 227–55.

LAWTON, R. (1968) Population changes in England and Wales in the later nineteenth century: an analysis of trends by registration districts, *Transactions Institute of British Geographers* **44**, 55–74.

PINCHEMEL, P. (1957) *Structures sociales et dépopulation rurale de la plaine picarde de 1836 à 1936*, Armand Colin, Paris.

PINCHEMEL, P. (1969) *France: a geographical survey*, Bell, London.

SHEPPARD, J. A. (1961) East Yorkshire's agricultural labour force in the mid-nineteenth century, *Agricultural History Review* **9**, 43–54.

SHEPPARD, J. A. (1962) Rural population change since 1851: three sample studies, *Sociological Review* **10**, 81–95.

Contemporary or near contemporary views of the decline of the agricultural labour force in nineteenth-century Britain are found in:

BOWLEY, A. (1914) Rural population in England and Wales: a study of the changes of density, occupation and ages, *Journal of the Royal Statistical Society* **77**, 597–652.

EVERSLEY, L. (1907) The decline in the number of agricultural labourers in Great Britain, *Journal of the Royal Statistical Society* **70**, 267–319.

LONGSTAFF, G. (1893) Rural depopulation, *Journal of the Royal Statistical Society* **56**, 380–442.

RAVENSTEIN, E. G. (1885) The laws of migration, *Journal of the Royal Statistical Society* **48**, 167–235.

Analyses of the reasons for recent reductions in the agricultural labour force are presented by:

AHLBERG, G. (1956) Population trends in Sweden, 1911–1950, *Lund Studies in Geography*, Series B, 16.

COWIE, W. and GILES, A. (1957) An enquiry into reasons for the "drift from the land", *Selected Papers in Agricultural Economics: University of Bristol* **5**, 70–113.

Community, or local, reasons for rural outmigration are considered in:

BRACEY, H. E. (1958) A note on rural depopulation and social provision, *Sociological Review* **6**, 67–74.

BRACEY, H. E. (1959) *English Rural Life*, Routledge & Kegan Paul, London.

MITCHELL, G. D. (1950) Depopulation and rural social structure, *Sociological Review* **42,** 69–85.
MITCHELL, G. D. (1951) The relevance of group dynamics to rural planning problems, *Sociological Review* **43,** 1–16.
RUTMAN, G. L. (1970) Migration and economic opportunities in West Virginia, *Rural Sociology* **35,** 206–17.
THORNS, D. C. (1968) The changing system of social stratification, *Sociologia Ruralis* **8,** 161–78.

Personal reasons for outmigration and the selectivity of migration are considered in:

ASHBY, A. W. (1939) The effects of urban growth on the countryside, *Sociological Review* **31,** 345–69.
HANNAN, D. F. (1969) Migration motives and migration differentials among Irish rural youth, *Sociologia Ruralis* **9,** 195–220.
HOUSE, J. W. (1965) *Rural North-east England: 1951–61*, Department of Geography, University of Newcastle-upon-Tyne, Papers on Migration and Mobility, **1**.
JANSEN, C. (1969) Some sociological aspects of migration, in JACKSON, J. A. (ed.), *Migration*, Cambridge University Press, London.
JONES, H. (1965) A study of rural migration in central Wales, *Transactions Institute of British Geographers* **37,** 31–45.
MARTINSON, F. M. (1955) Personal adjustment and rural/urban migration, *Rural Sociology* **20,** 102–10.
RAMBAUD, P. (1969) *Société rurale et urbanisation*, Seuil, Paris.
WAKELEY, R. E. and NASRAT, M. E. (1961) Sociological analysis of population migration, *Rural Sociology* **26,** 15–23.

Migration systems are described in:

BROWN, J. S. *et al.* (1963) Kentucky mountain migration and the stem-family: an American variation on a theme by Le Play, *Rural Sociology* **28,** 48–69.
CRAWFORD, C. O. (1966) Family attachment, family support for migration and migration plans for young people, *Rural Sociology* **31,** 293–300.
HILLERY, G. A. *et al.* (1965) Migrational systems of the southern Appalachians: some demographic observations, *Rural Sociology* **30,** 33–48.
KANT, E. (1951) Studies in rural/urban interaction, *Lund Studies in Geography*, Series B, **3,** 3–13.
PINKNEY, D. H. (1953) Migrations to Paris during the Second Empire, *Journal of Modern History* **25,** 1–12.

The behavioural approach to migration and other human activities is considered in:

MORRILL, R. L. and PITTS, F. (1967) Marriage, migration and mean information field: a study in uniqueness and generality, *Annals Association of American Geographers* **57,** 401–22.
PRED, A. (1967) Behavior and location, *Lund Studies in Geography*, Series B, **27**.
WOLPERT, J. (1964) The decision process in the spatial context, *Annals Association of American Geographers* **54,** 537–58.
WOLPERT, J. (1965) Behavioral aspects of the decision to migrate, *Papers of the Regional Science Association* **15,** 159–67.

A valuable theoretical approach to the study of rural/urban migration is found in:

MABOGUNJE, A. (1970) Systems approach to a theory of rural/urban migration, *Geographical Analysis* **2,** 1–15.

Models of migration are summarized in:

HAGGETT, P. (1965) *Locational Analysis in Human Geography*, Edward Arnold, London.
MORRILL, R. L. (1965) Migration and the spread and growth of urban settlement, *Lund Studies in Geography*, Series B, **26**.

Patterns of rural outmigration are considered by:

FRIEDLANDER, D. and ROSHIER, R. J. (1966) A study of internal migration in England and Wales, *Population Studies* **19,** 239–80.

HANNERBERG, D. (ed.) (1957) Migration in Sweden: a symposium, *Lund Studies in Geography*, Series B, **13.**

HOOKWAY, R. J. S. (1958) A study of rural population structure, *Journal of the Town Planning Institute* **44,** 210–13.

WENDEL, B. (1953) A migration schema: theories and observations, *Lund Studies in Geography*, Series B. **9.**

WILLATTS, E. J. and NEWSON, M. G. (1953) The geographical pattern of population changes in England and Wales, 1921–51, *Geographical Journal* **119,** 431–55.

ZELINSKY, W. (1962) Changes in the geographic patterns of rural population in the USA, *Geographical Review* **52,** 492–524.

CHAPTER 3

PEOPLE IN THE COUNTRYSIDE

SOCIAL STUDIES OF RURAL LOCALITIES

Many studies of social life at various localities in the British countryside have been prepared by social anthropologists, sociologists, and human geographers. Usually these have been isolated investigations describing local conditions in detail. However, the contents of these monographs may be placed in a broader context to portray situations which, in varying degrees, are different from life in towns. It is not the intention of this chapter to rehearse the specific contents of such locality studies. A valuable summary exists in the work of R. Frankenberg (1966). It would not be possible to do justice to the wealth of information in these various rural monographs in just one short chapter. Instead, an attempt will be made to isolate the distinguishing characteristics of rural life and to illustrate them by brief reference to selected locality studies.

Numerous studies by social scientists have recognized differences between life in towns and in the countryside. Perhaps the work of L. Wirth (1938), entitled *Urbanism as a Way of Life*, has been the most influential. This outlined the theory that an increase in the size and density of population in a settlement led to increased anonymity, a more widespread division of labour, and, in turn, greater social heterogeneity. Relationships became impersonal so that the urbanite lost the "spontaneous self expression, morale and sense of participation" that came from living in an integrated, rural society (p. 54). City living was therefore quite different from country living.

More recently A. D. Rees (1951) presented a spirited version of this opinion in his defence of rural life: "The failure of the urban world to give its inhabitants status and significance in a functioning society, and their consequent disintegration into formless masses of rootless nonentities, should make us humble in planning a new life for the countryside. The completeness of the traditional rural society—involving the cohesion of family, kindred and neighbours—and its capacity to give the individual a sense of belonging, are phenomena that might well be pondered by all who seek a better social order" (p. 170). N. Anderson (1963) has described the existence of a similar bias in the United States. "The American ideal of home, community and work was clearly rural and puritanic. Industrialism and urbanism combined in various ways to shock all established norms. . . . The city came to be associated with every vice and evil and these attitudes of bias came to be rooted in most established institutions" (p. 8).

In the past, the qualification "rural" applied to areas of low population density, with settlements of small absolute size located in relative isolation to their surroundings. Farming formed the major economic base in such localities, where the way of life was reasonably homogeneous and different from that of other sectors of society, namely the "city". P. Sorokin and C. Zimmermann (1929) spelled out eight groups of variables which they considered would distinguish rural from urban conditions: occupation, environment, community size, population density, homogeneity of the population, social differentiation, mobility, and systems of interaction. The "rural" characteristics of each variable will be described in the sentences which follow.

(i) Rural localities contained a high proportion of workers and their families directly occupied with managing land for farming and forestry. Nevertheless, such localities also included craftsmen and workers in essential local services before the process of ruralization set in.

(ii) The geographical environment of rural areas was considered to be predominantly natural rather than man-made, with the landscape being made up of fields and woodland rather than buildings, factories, and streets. Such a differentiation of land use is viable but, as is well known to geographers if not to other social scientists, most landscapes in the countryside are far from being natural but are modified by man, if not being actually man-made.

(iii) Rural settlements are normally smaller than towns. This general statement is acceptable, but it is not possible to offer precise size limits that have universal applicability to distinguish rural localities from urban ones.

(iv) Population densities are lower in country areas than in towns.

(v) Compared with urban dwellers, the populations of rural communities are more homogeneous in their social traits. Characteristics linked to language, beliefs, opinions, *mores*, and patterns of behaviour tend to be more uniform in the countryside than in towns. Religious and family institutions survive more readily in the countryside. This point was certainly valid in an historic context but is less so in an age of mass media of communication.

(vi) Social differentiation and stratification are more evident in the town than in the countryside. Class differences are less pronounced. P. Mann (1965) has shown that "the class system of the urban environment, based greatly upon secondary social contacts does not operate in the same way in the village. The village does not have its different types of residential areas to anything like the degree found in the city. There are not large aggregates of professional men, . . . industrial workers and so on" (p. 18). Occupational differentiation certainly exists in rural areas, but the argument runs that face-to-face contacts provide a sense of community-belonging to counterbalance social differentiation.

(vii) Mobility, in both a spatial and a social sense, is less intense in the countryside than in the town. Nevertheless, work by W. M. Williams (1963) in England and T. Hägerstrand (1957) in Sweden has shown that important short-distance movements took place between neighbouring settlements before the railway age, involving, among others, farming families in search of holdings to suit their position in the family cycle. But this group was less mobile than craftsmen and farm labourers, as Table 6, derived from Williams' work, suggests. These two categories

TABLE 6. CHANGES IN RESIDENCE OF FAMILIES IN ASHWORTHY PARISH, 1851, COMPARED WITH 1841 (BY SOCIO-ECONOMIC STATUS OF HEAD OF HOUSEHOLD)

	Farmers (%)	Craftsmen (%)	Farm labourers (%)
Occupying same house (farm*)	73.3	54.7	37.4
Changed house (farm*) within parish	15.2	24.7	30.8
Moved into Ashworthy parish	8.2	20.5	31.8

* In the case of farmers.

Source: WILLIAMS, W. M. (1963), *A West Country Village: Ashworthy—family, kinship and land*, Routledge & Kegan Paul, London, p. 128.

N.B.—Totals less than 100 per cent due to changes in the status of some residents.

contained important proportions of unmarried people who had no fixed stake in either land or equipment and often had to change their place of residence when their contracts terminated.

(viii) By virtue of living in small settlements with a relative lack of transportation to permit them to move around easily and frequently to other settlements, rural dwellers in the past had far fewer human contacts than had urban dwellers. But rural relations were at a personal, face-to-face level. The rural economy might have been simple, but this itself engendered a complexity in social life. In truly rural societies the social network was close-knit, with everybody knowing and interacting with everyone else. In urban society individuals had few friends in common. Interactions in rural society splayed across the following five areas of activity: kinship, economic life, politics, ritual or religious activity, and recreation.

R. Frankenberg (1966) explained that in the past inhabitants of the countryside were members of distinguishable communities. In other words, residents of rural localities "have overriding economic interests which are the same or are complementary. They work together and also play and pray together. Their common interest in things gives them a common interest in each other" (p. 238). There has been much controversy about the precise meaning of the concept of community. However, G. A. Hillery (1955) found that most social scientists were "in basic agreement that 'community' consists of persons in social interaction within a geographical area and having one or more common ties" (p. 111). In the past communal systems of agricultural organization operated in many regions, involving not only ploughing, seeding, and harvesting, but also stock-grazing, sheep-dipping, and sheep-shearing, often using communal implements. In addition to such agricultural activities, H. Mendras (1965) has emphasized the historic importance of communal responsibilities for maintaining roads and other parts of the fabric of the countryside in Continental Europe.

Many such systems have now disappeared, but remnants of past obligations of mutual help in agricultural tasks can be detected. Thus in western Ireland in the 1930's the inhabitants treated "cooperation amongst farmers in economic tasks in the same way as the obligations of kin to help at weddings and at funerals, or to give stock and services to mitigate disaster. . . . The man who refuses to help, or for that matter to be helped, is opting out of society and condemning himself to social isolation. The man who helps his neighbour and kinsman and is helped in his turn—is, with each exchange of services, cementing and reinforcing the ties which bind him to the community. He strengthens the strands of the network, the existence of which enables us to speak of Irish townlands as communities" (Frankenberg, 1966, p. 37).

The work of C. Arensberg and S. Kimball in western Ireland exemplified how in a community of economic identity of interest—such as existed among the small farmers—economic and social life, politics and kinship, were inseparably linked together. Other rural studies in western Britain emphasized the historic importance of mutual aid. T. Jones-Hughes (1960) explained that "the basis of these obligations of mutual aid in north Wales is to be sought in the nature of the family farm as a corporate working group. The family as an ideal working unit is complete only for a short period of time in its history, that is, when both parents are alive and active and when the children are old enough to make a contribution towards the running of the farm without necessarily receiving any regular cash remuneration" (p. 161). But there would be long periods in the family cycle when outside help was needed from neighbours and kin during particularly busy periods of the agricultural year. W. M. Williams (1953) in his study of family farming in Cumberland demonstrated how historic problems of co-operation amongst farmers were strongly linked to the kinship system and also how these social contacts weakened as agricultural mechanization reduced the need for mutual aid in farming practices.

CHANGES IN RURAL/URBAN DIFFERENCES

Some social scientists have recognized differing degrees of rurality or urbanity at specific localities and have placed such localities (sometimes referred to as communities) along a graded scale between two polarities which might be labelled "truly rural" and "truly urban". T. L. Smith (1951) explained the reasons for such an arrangement: "Rural and urban do not exist of themselves in a vacuum, as it were, but the principle characteristics of each may be found shading into, blending or mixing with the essential characteristics of the other. . . . Rather than consisting of mutually exclusive categories, rural and urban, the general society seems to resemble a spectrum in which the most remote backwoods sub-rural settlements blend imperceptibly into the rural and then gradually through all degrees of rural and suburban into the most urban and hyperurban way of living. If such be the case, a scale, rather than a dichotomy, would provide the most satisfactory device for classifying the population or the groups according to rural or urban characteristics."

Such an approach was used by R. Frankenberg (1966) when he arranged locality studies in the British Isles along a morphological continuum in his book entitled *Communities in Britain*. Differences between social conditions at localities in the countryside reported in the book exemplified variations in the strength of the kinds of "community" features that have already been outlined. However, many parts of Britain were undoubtedly rural in spatial or visual terms, measured by land-use conditions, but contained very few social characteristics which pointed to the existence of a community. The study of Westrigg parish in the Cheviots by J. Littlejohn (1963), unlike many other locality investigations, did not involve family farming but described vast sheep ranches of hundreds or thousands of hectares owned by absentee landlords and run by shepherds and farm labourers. Kinship links were far less significant in Westrigg than in the other localities investigated elsewhere in Britain. "Family" was recognized only in terms of the nuclear group of parents and children rather than in the context of the extended family. In Littlejohn's words: "most parishioners live in Westrigg not because the parish forms a group they have to belong to or in which they have special rights, but because their occupation or that of their husband or father lies there" (p. 10). With the passage of time "the parish has become less and less a social unit possessing an independent existence of its own, and parishioners have been drawn increasingly into a wider network of contacts and relationships" (p. 39). Littlejohn's study showed that the British farmworker's family now has little or nothing to do with the farm. Their life is quite as separate from the man's work as are the lives of family members of industrial employees. In Westrigg an awareness of social class seemed to have become more noticeable than an appreciation of community. The analysis suggested strong social similarities between those who worked in towns and those who worked in the countryside.

The traditionally recognized contrast between rural and urban conditions and the concept of a linking continuum have been reviewed critically by R. E. Pahl (1966). He emphasized that whilst rural areas could be readily distinguished from urban ones when land-use criteria were examined, this was not the case when social variables were considered. In his own words: "In the sociological context the terms rural and urban are more remarkable for their ability to confuse than for their power to illuminate" (p. 299). A review of international evidence showed that there were some striking exceptions to the classic rural and urban environments outlined by Wirth.

On one hand, Pahl found much evidence to show that the social organization of the inhabitants of the central areas of some cities were quite different from the "impersonal, superficial, transitory and segmental" social environments described by Wirth (p. 54). Face-to-face relationships and kinship links remained strong in "urban villages" which could be identified in city areas from the East End of London, to Boston, Delhi, Mexico City, and doubtless elsewhere. The existence of such situations was in opposition to traditional views of urban life.

On the other hand, important social changes were taking place in "metropolitan villages" which housed increasing numbers of middle-class, often city-born, commuters around large cities in the developed world and thereby changed the social composition of what had become pseudo-rural settlements. For example, Pahl was able to identify the following groups of residents in "metropolitan villages" north of London: large property owners; the salariat; retired urban workers with some capital; urban workers with limited incomes (the "reluctant commuters" forced out of the city by high property prices); rural working-class commuters who had houses in the village but were obliged to seek jobs elsewhere; and the traditional ruralities, including agricultural workers and tradesmen whose employment as well as residence was local. Thus a "dispersed city" was being created, composed of new towns, old towns, villages, and hamlets beyond the built-up limits of central cities, but increasingly characterized by middle-class inhabitants with urban life styles and points of view. Interestingly enough, L. Wirth had suggested back in the 1930's that urban conditions were not necessarily those *in* the city but those *of* the city.

Pahl examined a number of non-European studies which illustrated a variety of interrelationships between inhabitants of town and country areas. He concluded that: "rural communities can be isolated as separate systems for the purpose of academic study, but this is an increasingly unreal exercise" (p. 316). Instead, emphasis should be placed on the outlooks and attitudes of groups and individuals. He explained "the notion of a rural/urban continuum arose in reaction against the polar-type dichotomies, but there are equal dangers in over-readily accepting a false continuity. Not only is there a whole series of continua but there are sharp discontinuities, in particular between the local and the national . . . it is clear it is not so much *communities* that are acted upon as groups and individuals at particular places in the social structure. Any attempts to tie particular patterns of social relationships to specific geographical milieux is a singularly fruitless exercise" (p. 322).

Rather than identifying a single continuum from the "truly rural" to the "truly urban", Pahl argued that "it would be better to imagine a whole series of meshes of different textures superimposed on each other, together forming a proces' which is creating a much more complex pattern" (p. 327). Whilst accepting this most important point that *individuals* are affected by the forces of social change rather than all inhabitants at particular geographical localities, it is relevant to note that certain parts of the countryside are more likely to be modified by urban influences than others. Such areas include not only commuting hinterlands around central cities, where metropolitan villages have developed, but also areas which contain large numbers of second homes for weekend and/or holiday use, and other parts of the country which are favoured by tourists or retired people.

The various locality studies produced in the British Isles, and summarized by Frankenberg and others, relate to conditions in the past, namely from the 1930's onwards. In 1951 A. D. Rees could note that in central Wales "the countryman has continued to live in a world of his own, the standards of which differ from those of our modern industrial civilization" (p. 30). This comment has decreasing validity with every month that passes. In H. E. Bracey's (1970) words "the village community as it has been understood in the past, functioning as a separate entity, is rapidly disappearing. . . . The fact is that during the 1960's changes have been so rapid and so profound that published social surveys do not reflect the 1970 way of living" (p. 137).

P. Mann (1965) argued that "with current developments of both technological and social kinds the value of the rural/urban dichotomy and the linking continuum is being seriously challenged as a rather out-dated tool which might have been of value when . . . in England few village dwellers ever went beyond the nearest market town. But today, with changing conditions, it seems to be losing, if it has not already lost, its value as an analytical tool" (p. 71).

J. A. Beegle (1964) reported that enormous social changes have already taken place amongst farmers in North America. Similar modifications may be expected to affect agriculturalists to varying degrees in other parts of the developed world in the future. In North America "the farmer is a business man who keeps a sharp eye on domestic and world markets, applies scientific methods in seeding and feeding, owns a car and TV set, and has his wife and daughter dressed according to the latest fashion. . . . Ecologically speaking, the American farmer does not live in a city, yet his ways are citified. He is *of* the city even though he is not *in* the city" (p. 242). He forms "an organic segment of the national economy, and there is an unprecedented degree of interdependence with the national economy. Moreover, the farmer is involved in a process of desegregation as to the social institutions in which he participates. The open country has been invaded by non-farm people and the level of living of farm people is approaching that of other segments" (Nelson, 1957, p. 20). Differences in attitudes and value systems held by farmers, on the one hand, and by other sections of the population in North America, on the other, have not disappeared entirely but have been reduced dramatically since 1945.

In France the old social structure of rural areas has changed radically over the same time period. Previously, three main groups could be recognized in most regions: a dominant group of large landowners; a dominated group of small farmers, agricultural labourers, and rural craftsmen; and a relatively independent group of semi-subsistence agriculturalists (George, 1964, 1965). Following depopulation, this type of agricultural society has disappeared almost entirely from progressive parts of the countryside, such as the Paris Basin. It has been replaced by city-orientated country dwellers made up of farm owners and managers, skilled farm technicians, and service personnel in transport, commerce, and other activities.

G. P. Wibberley (1960) has summarized the current ambiguity of the terms rural and urban. "In the middle of the twentieth century we are uncertain as to what is really meant by the term 'rural community', and whether there are now any significant differences between rural and urban people in the life they lead, in their hopes and aspirations, and in their attitudes and *mores*. . . . The dominant cause of recent changes in the structure and function of rural society has been the breakdown of self-sufficiency in country areas through the phenomenal growth of material and personal mobility. . . . The problem for most European countries is to achieve an environment, even in their thinly populated rural areas, where persons with urbanized minds can live and work happily" (p. 121).

The chapters which follow will outline: first, the implications of increased personal mobility and the various social changes that have taken place in areas that are still visually recognizable as "countryside", and, second, some of the management problems which have to be tackled in country areas which are experiencing new and changing pressures on their local resources.

REFERENCES AND FURTHER READING

Introductions to socio-geographical and sociological studies are provided by:

PAHL, R. E. (1965) Trends in social geography, in CHORLEY, R. J. and HAGGETT, P. (eds.), *Frontiers in Geographical Teaching*, Methuen, London, pp. 81–100.

PAHL, R. E. (1967) Sociological models in geography, in CHORLEY, R. J. and HAGGETT, P. (eds.), *Models in Geography*, Methuen, London, pp. 217–42.

A summary of some locality studies in the British Isles is provided by:

FRANKENBERG, R. (1966) *Communities in Britain*, Penguin, Harmondsworth.

A critical comment on this summary is found in:

PAHL, R. E. (1970) *Patterns of Urban Life*, Longmans, London.

The substance and methods of community studies are considered in:

BELL, C. and NEWBY H. (1971) *Community Studies: an introduction to the sociology of the local community*, George Allen & Unwin, London.

An important rural locality study is by:

WILLIAMS, W. M. (1963) *A West Country Village: Ashworthy—family, kinship and land*, Routledge & Kegan Paul, London.

Other rural locality studies by anthropologists, geographers and sociologists include:

ARENSBERG, C. M. and KIMBALL, S. T. (1940) *Family and Community in Ireland*, Peter Smith, London.

CRESSWELL, R. (1969) *Une Communauté Rurale d'Irlande*, Travaux et Memoires de l'Institut d'Ethnologie, Paris.

DAVIES, E. and REES, A. D. (eds.) (1960) *Welsh Rural Communities*, University of Wales Press, Cardiff.

EMMETT, I. (1964) *A North Wales Village*, Routledge & Kegan Paul, London.

JONES-HUGHES, T. (1960) Aberdaron: the social geography of a small region in the Llŷn peninsula, in DAVIES, E. and REES, A. D. (eds.), *Welsh Rural Communities*, University of Wales Press, Cardiff, pp. 121–81.

HAGERSTRAND, T. (1957) Migration and area, *Lund Studies in Geography*, Series B, **13**, 27–158.

LEWIS, G. J. (1970) A Welsh rural community in transition: a case study in mid-Wales, *Sociologia Ruralis* **10**, 143–61.

LITTLEJOHN, J. (1963) *Westrigg: the sociology of a Cheviot parish*, Routledge & Kegan Paul, London.

MARTIN, E. (1965) *The Shearers and the Shorn: a study of life in a Devon community*, Routledge & Kegan Paul, London.

NALSON, J. S. (1968) *The Mobility of Farm Families*, Manchester University Press, Manchester.

REES, A. D. (1951) *Life in a Welsh Countryside*, University of Wales Press, Cardiff.

THABAULT, R. (1971) *Education and Change in a Village Community: Mazières-en-Gâtine*, 1848–1914 Schoken Books, New York.

WILLIAMS, W. M. (1953) Some social aspects of recent changes in agriculture in west Cumberland, *Sociological Review* **1**, 93–100.

WILLIAMS, W. M. (1956) *The Sociology of an English Village: Gosforth*, Routledge & Kegan Paul, London.

WILLIAMS, W. M. (1958) *The Country Craftsman*, Routledge & Kegan Paul, London.

WILLIAMS, W. M. (1963) The social study of family farming, *Geographical Journal* **129**, 63–75.

WILLIAMS, W. M. (1964) Changing functions of the community, *Sociologia Ruralis* **4**, 299–310.

WYLIE, L. (ed.) *Chanzeaus: a village in Anjou*, Harvard University Press, Cambridge, Mass.

The concept of "community" and the status of community studies are discussed by:

FREILICH, M. (1963) Towards an operational definition of 'community', *Rural Sociology* **28,** 117–27.
HILLERY, G. A. (1955) Definitions of community: areas of agreement, *Rural Sociology* **20,** 111–23.
HILLERY, G. A. (1961) The folk village: a comparative analysis, *Rural Sociology* **26,** 337–53.
SIMPSON, R. L. (1965) Sociology of the community: current status and prospects, *Rural Sociology* **30,** 127–49.
SUTTON, W. A. and KOJALA, J. (1960) The concept of "community", *Rural Sociology* **25,** 197–203.

The rural/urban continuum and rural/urban differences are considered in:

ANDERSON, N. L. (1963) Aspects of the rural and urban, *Sociologia Ruralis* **3,** 8–22.
BEALER, R. C. *et al.* (1965) The meaning of "rurality" in American society, *Rural Sociology* **30,** 255–66.
BEEGLE, J. A. (1964) Population changes and their relationship to changes in social structure, *Sociologia Ruralis* **4,** 238–52.
BEEGLE, J. A. (1966) Social structure and changing fertility of the farm population, *Rural Sociology* **31,** 415–27.
BENET, F. (1963–4) Sociology uncertain: the ideology of the rural–urban continuum, *Comparative Studies in Society and History* **6,** 1–23.
FUGUITT, G. V. (1963) The city and the countryside, *Rural Sociology* **28,** 246–61.
MACGREGOR, M. (1972) The rural culture, *New Society*, 9 March 1972, 486–9.

Recreational travel is considered in:

MANN, P. H. (1965) *An Approach to Urban Sociology*, Routledge & Kegan Paul, London.
PAHL, R. E. (1966) The rural/urban continuum, *Sociologia Ruralis* **6,** 299–329; also in PAHL, R. E., (ed.), *Readings in Urban Sociology*, 1968, Pergamon, Oxford, pp. 263–97.
PAHL, R. E. (1967) The rural/urban variable reconsidered: the cross-cultural perspective, *Sociologia Ruralis* **7,** 21–30.
SCHNORE, L. F. (1966) The rural/urban variable: an urbanite's perspective, *Rural Sociology* **31** 415–27.
SOROKIN, P. and ZIMMERMANN, C. (1929) *Principles of Rural–Urban Sociology*, New York.
STREIB, G. F. (1970) Farmers and urbanites: attitudes toward intergenerational relations in Ireland, *Rural Sociology* **31,** 346–54.
WIRTH, L. (1938) Urbanism as a way of life, *American Journal of Sociology* **44,** 46–63.

See also the rejoinders to the article by Pahl (1966):

LUPRI, E. (1967) The rural/urban variable reconsidered: the cross-cultural perspective, *Sociologia Ruralis* **7,** 1–20.
OOMMEN, T. K. (1967) The rural/urban continuum re-examined in the Indian context, *Sociologia Ruralis* **7,** 30–48.
SMITH, T. L. (1951) The rural and rural worlds, in SMITH, T. L. and MCMAHAN, C. A. (eds.), *The Sociology of Urban Life*, New York.

Recent social changes in rural areas are considered in:

BRACEY, H. E. (1970) *English Rural Life*, Routledge & Kegan Paul, London.
GEORGE, P. (1964) Anciennes et nouvelles classes sociales dans la campagne française, *Cahiers Internationaux de Sociologie* **37,** 13–21.
GEORGE, P. (1965) Quelques types régionaux de composition sociale dans les campagnes françaises, *Cahiers Internationaux de Sociologie* **38,** 49–56.
MENDRAS, H. (1965) *Sociologie de la Campagne Française*, Presses Universitaires de France, Paris.
MENDRAS, H. (1970) *La Fin des Paysans*, Armand Colin, Paris.
NELSON, L. (1957) Rural life in a mass industrial society, *Rural Sociology* **22,** 20–30.
WALLACE, D. B. (1967) Some social considerations, in ASHTON, J. and ROGERS, S. J. (eds.), *Economic Change and Agriculture*, Oliver & Boyd, Edinburgh.
WIBBERLEY, G. P. (1960) Changes in the structure and function of the rural community, *Sociologia Ruralis* **1,** 118–27.

CHAPTER 4

URBANIZATION
OF THE COUNTRYSIDE—I

OVERVIEW

In the preceding discussion attention has been drawn to the retreat of population from the countryside as it moved by preference to towns and cities. The reverse of this trend has, however, become apparent since World War II as city dwellers in many developed countries have moved into the countryside in search of homes and recreation and have thereby produced more and more examples of what R. E. Pahl (1965b) has called "mentally urbanized, but physically rural parts of the country" (p. 5). In G. P. Wibberley's (1960) words: "in highly developed countries a common culture is arising between town and country" (p. 121). Even before World War II this had been anticipated in Britain by A. W. Ashby (1939) who insisted that "slowly but surely, yet much more rapidly than formerly, the rural population will absorb most of the elements of the common culture, and the adoption of common standards throughout will be for the good of society as a whole" (p. 369).

The urbanization of the countryside can be produced by a variety of mechanisms and take on a number of nuances of which four have been selected for discussion here. Increasing affluence, efficient public transportation, and rising rates of private-car ownership have combined to allow ever-growing numbers of city people to realize their choice to live in parts of the country that are still visually rural and yet also to commute into town for work, education, entertainment, and the acquisition of goods and services. This is a well-established phenomenon in many parts of the developed world which will be illustrated in the context of Great Britain in the following discussion.

A variation on this theme is provided by the "worker-peasant"* phenomenon which is found in many parts of Continental Europe. Unlike the preceding form of urbanization, which involved city people choosing to live in the country and work in the town, the worker-peasant phenomenon involves members of farming families commuting to city jobs but still living in their farmhouses and working their holdings on a part-time basis. This kind of contact between farming people and urban/industrial environments has been investigated in a number of detailed case studies undertaken in Continental Europe. The section which follows will therefore be cast in that geographical context.

* Usual translation of *ouvrier-paysan* (Fr.) or *Arbeiter-Bauer* (Ger.).

The movement of city people into the countryside for recreation, both during vacations and over weekends, provides another illustration of the diffusion of urban ideas and values which provokes changes in the attitudes and outlooks of country people. French rural sociologists have conducted detailed studies of such contacts in Alpine ski resorts, and their experience will serve as a basis for discussion.

Finally, the trend for city people to acquire cottages or second homes in the countryside for leisure use has occurred in many parts of the developed world. Sufficient material has not yet been published to be able to consider the implications of the presence of second homes in Britain, but information will be drawn from numerous studies which have been prepared in other countries.

COMMUTER SETTLEMENTS

The Dispersed City

Rising rates of car ownership since 1950 have permitted the dispersal of the city into the countryside in Great Britain and have thereby transformed town/country relations. The change has been so great that in Ruth Glass's (1962) words: "the countryside is overrun and festooned with ribbons of pseudo-rural habitations" (p. 487). As a result "the bogy of rural depopulation which was with us in Great Britain for a whole century has been banished: most rural areas within 30 km of a sizeable town now report increases in their resident population" (Bracey, 1963, p. 75). A number of case studies in peri-urban areas have been undertaken, and their findings will form the substance of the following discussion. To take a single example, the village of Drymen, 27 km north of Glasgow, underwent a long period of population decline between 1850 and 1951 but in the last two decades the trend has been reversed as the commuters' invasion has occurred (Green, 1964).

Figure 4.1 depicts the generalized commuting hinterlands around major British cities as revealed by the findings of the 1961 census. The bounding lines circumscribe all component administrative areas which dispatched at least 100 workers to the central city (Lawton, 1968). As one would expect, the intensity of commuting becomes greater in the more immediate proximity of the city in question. However, there are important differences between individual villages in these commuter zones, with villages along main lines of communication, such as railways, main roads, and motorways, contrasting with settlements away from such axes which remain far more rural and agricultural in character. Masser and Stroud (1965), in their study of commuter settlements in the Wirral, concluded that there was little apparent correlation between linear distance from an urban area and the extent to which villages satisfied concepts of rural character and communal identity and were chosen for residence by commuters.

In addition to accessibility, a wide range of complicating factors has to be recognized to help explain the observed distribution of commuter villages around urban centres. These factors include: zoning regulations (especially the existence of green belts in Great Britain and other planning constraints); policies implemented by local authorities for expanding some settlements by permitting the

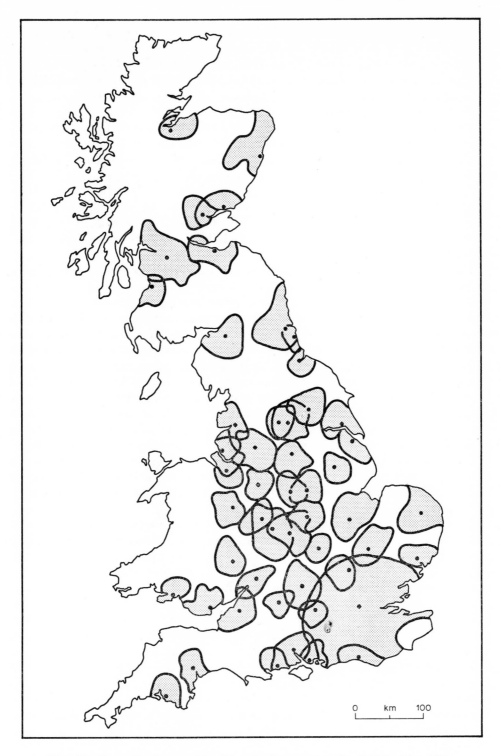

Fig. 4.1. Generalized commuting hinterlands around major British cities, 1961.

development of estates of private housing and/or council housing; the fact that some villages have been selected by commuters because of their reputation of "fashionability"; and, finally, spatial variations in property ownership which conditioned the release of land for development. D. Kendall's (1963) portrait of a disappearing English village in the southern Midlands provides a useful hand specimen of how the policies of landowners can limit, or indeed prevent, the expansion of rural settlements.

Commuting hinterlands become broader with every year that passes. A recent article has predicted that by the end of the 1970's there will be no more country houses "readily available within striking distance of any major city" in Britain (*The Times*, 25 October 1971). The motorway building programme has increased the distance that commuters are willing to travel; what used to be considered the 65 km "London barrier" is moving out all the time. G. P. Wibberley (1960) was undoubtedly right when he noted that "the thinning out of the urban mass, together with its increase in area, seems to have no natural or automatic limit" (p. 122).

The dispersed city has spread out from central cities via new towns, expanded market towns, villages, and hamlets which now combine to form city regions. Old cottages have been converted for modern living and new estates of public- and local-authority housing have been constructed together with isolated residences. As a result, the "settlement pattern which has existed in Great Britain for centuries is being assailed on every hand" (Bracey, 1963, p. 76). Local service facilities have not disappeared from such invaded settlements by contrast with changes in more remote parts of the country. Instead, they have been modified to serve the newcomers. Villages are thus being transformed by the presence of commuters, and development has tended to destroy the character of such settlements. "They are no longer the sorts of place where a country person can feel at home. The talk in the bar tends to be about production figures instead of harvest yields" (*The Times*, 25 October 1971). As a result, it might be argued that in less than a quarter of a century since Britain entered the motorway age the country will have experienced possibly the greatest social upheaval since the Industrial Revolution.

Changes in working and living habits have been so rapid in recent years that distinctions between town and countryside and between urban and rural communities are becoming very blurred. The edge of the city can certainly be recognized in purely formal terms as the line where continuously built-up land gives way to other, more open forms of land use. But such a definition cannot be translated directly into either functional or social terms. R. E. Pahl (1965a) commented that the geographer's understanding of the metropolitan fringe has all too frequently been handicapped because of his obsession with land use, so that for him an urban area tended to be defined in terms of bricks and mortar. Many residents of localities way beyond the built-up city edge are *of* the city if not *in* the city. In short, as Pahl (1966) noted: "in the sociological context, the terms rural and urban are more remarkable for their ability to confuse than for their power to illuminate" (p. 299). After reviewing various definitions of the rural/urban fringe put forward by R. A. Kurtz and J. B. Eicher (1958) he concluded that "the fringe is defined in relation to the city and spatially exists in the agricultural hinterland where land use is changing. The population density is increasing rapidly and land values are rising. Ecologically it could be said to be an area of invasion" (p. 74).

It is clearly no longer satisfactory to assume that officially defined *rural districts* in England and Wales (or, indeed, their official equivalents in other parts of the world) correspond with "the countryside". Isobel Robertson's (1961) analysis of the occupational structure of the inhabitants of *rural districts* in England and Wales derived from the 1951 census, distinguished three grades of district (Fig. 4.2). The first involved rural districts which were "agricultural/rural" with more than half of their employed population working on the land in 1951. Such districts were found in parts of the following areas which are remote from large cities: the Fens, Devon, central Wales, East Anglia, the North Riding of Yorkshire, Cumberland, and the Welsh borderland. "Agricultural/rural" districts had a poor provision of commercial, social and educational services. Remoteness appeared to discourage the settling of adventitious population. The second type of district was the "rural" class, with between one-third and one-half of the employed population in farming. However, the final "rural/urban" type of district predominated, with less than one-third of the working population on the land. Such areas had high proportions of adventitious population, many of whom commuted to work in nearby urban centres. Three-quarters of all *rural districts* in England and Wales were already of this final pseudo-rural character in 1951. The proportion has surely increased since then.

One of the predominant features of settlements around major employment foci is their recent increase in population, principally as a result of inmigration by residents from nearby cities. The newcomers fall into two principal groups. The first comprises people moving into the country to retire. The second, and undoubtedly the larger, involves urbanites who can afford to choose to live in another environment and commute to work in the city each day. This group is composed essentially of young and middle-aged couples with children. The newcomers "are likely to be either fairly affluent professional people or manual workers coming to a particular job. The junior, non-manual, black-coated or white-collar worker tends to be barred from the countryside unless he is able to buy a cheap cottage or lives in a council house: the new privately built house is generally beyond his means" (Pahl, 1965a, p. 47).

Such people can afford to realize their dreams of rural living by purchasing existing property in the countryside or having architect-designed houses specially constructed. The spread of upper middle-class housing into rural areas along railway lines and motorways up to 80 km from central London has been described as the "cocktail belt". This is an "extreme manifestation of the spatial segregation of social classes which characterizes England more than any other European country" (Whitehand, 1967, p. 27). But less pretentious houses are, of course, occupied by the majority of newcomers.

In the American context: "the flight from the city may be a folk movement away from its dirt, violence, racial and religious tensions. The pastoral image of green fields, small community and basic primacy of family relationships may draw people away from the larger problems of the metropolis to the more manageable world

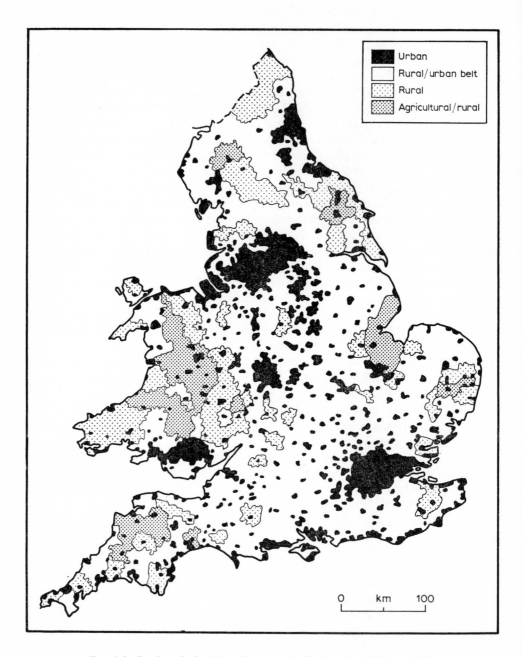

Fig. 4.2. Grades of administrative areas in England and Wales, 1951.

of the fringe" (Pahl, 1965a, p. 75). Such attractions of the countryside were certainly stressed as the advantages of village life by newcomers in commuter settlements in East Anglia (Table 7). The most frequently held opinion was that the village provided better surroundings for bringing up children. A large proportion of newcomers appreciated that it was easy to get to know people in a small settlement but, rather curiously, a sizeable number were keen to retain their urban anonymity and keep themselves to themselves.

TABLE 7. REPORTED ADVANTAGES OF VILLAGE
LIFE

	%*
Leisure activities	17
Easy to get to know the people	47
Can keep yourself to yourself	37
Cheap to live	22
Better surroundings to bring up children	64
Not available/don't know	13
	200

* Combination of first and second choices.

Source: EMERSON, A. E. and CROMPTON, R. (1968),
Report to the Suffolk Rural Community Council, Suffolk:
Some social trends. Mimeographed report, p. 39.

The application of development control in green-belt zones around Britain's major cities has diverted the growing demands for new housing, which arise from household formation and from urban redevelopment, from the built-up edge of the city so that they leapfrog into the visually rural hinterland. Pahl (1965b) has rightly concluded that "in an increasingly leisure-oriented society, space has become a status symbol, so that with increasing affluence the proportion of the population which wants, and can afford, to live in the country will grow" (p. 21). Whilst applying predominantly to the acquisition of country houses for full-time use, this demand is also being satisfied by the purchase of ever-growing numbers of second homes. This phenomenon will be discussed later.

Population growth in peri-urban areas results both from inmigration and from highish rates of natural increase. But a consideration of case studies shows that such patterns of net growth often conceal the continuing contraction of numbers of long-established residents as old forms of local employment are no longer available. Eleven villages close to Nottingham, studied by D. C. Thorns (1968), experienced both an absolute and a relative decline in their agricultural population whilst the number of commuters increased. G. J. Lewis's (1967) analysis of the social geography of the Welsh village of Bow Street demonstrated similar counterflows. During the nineteenth century the village had flourished because of local lead mining but by 1900 this source of prosperity had disappeared and the population had started to decline. Local sources of employment continued to contract during the first half of the present century as Bow Street's residents either migrated to jobs elsewhere or commuted on a daily basis to employment in the nearest town (Aberystwyth). Such an outflow of the established population has continued since World War II as young residents (essentially between 18 and 30 years of age) move away for education, at marriage, or for obtaining employment elsewhere. Important counterflows have developed over the past twenty years as Bow Street and other peri-urban settlements were selected as dormitory villages by middle-class folk employed in Aberystwyth. Sources of employment in settlements such as Bow Street have decreased in relative importance following the commuter invasion. In the early 1960's three-fifths of the employed population of Bow Street worked in Aberystwyth. The remainder was divided almost equally between Bow Street itself and its surrounding villages. Nevertheless, the net result of the newcomers' arrival in such settlements was to enlarge the resident population and thereby provide additional support for local shops, schools, and societies which might otherwise have closed. Such a beneficial effect was also confirmed in a study of social conditions in commuter villages in a part of East Anglia where "not only does the village continue to provide a social focus for the individual" but also [in the authors' opinion] "further development, at least on the social side, is possible" (Emerson and Crompton, 1968, p. 119).

Social Changes in British Commuter Settlements

Peri-urban zones are subject to major changes in land use, not simply in terms of accommodating new housing but also for the location of new facilities such as motorways and airports which essentially serve the urban centres. But in addition to such important land-use modifications, highly significant social changes have also taken place. R. E. Pahl (1965a), in his study of *Urbs in Rure*, has illustrated the movement of middle-class families into villages north of London where "the traditional world of a small, established middle class with a large working-class population has been invaded by a new middle-class commuting element so that now the middle-class group is numerically the greater" (p. 49).

Other case studies of commuter villages have demonstrated the recency of this middle-class invasion. In Pahl's sample villages in north-east Hertfordshire 81 per cent of the middle-class families (see Table 8 for definitions) had arrived between 1945 and 1961 by comparison with only 29 per cent of the working-class families arriving over the same period. The traditional rural social structure, arranged in an hierarchy (from the village squire, parson, and schoolmaster at the top, to the farm labourer at the base), has been modified and polarized into a more abrupt working-class/middle-class dichotomy in such settlements. The middle-class commuters who came in search of the rural idyll "try to get the 'cosiness' of village life, without suffering any of the deprivations, while maintaining a whole range of contacts outside" (Pahl, 1970, p. 97). As a result, the traditional social structure has been transformed to approximate with conditions elsewhere in urban Britain.

The behaviour patterns of members of the two classes were shown to be quite

TABLE 8. LENGTH OF RESIDENCE OF THE CLASSES

	"Established" (%)	"Newcomers" (%)
Middle class	19	81
Working class	71	29
Agricultural workers	54	46

Source: PAHL, R. E. (1965a), *Urbs in Rure*, London School of Economics Geographical Papers, **2,** 49.

N.B.—Middle class defined as heads of households in groups 1 and 2 in the Registrar-General's *Classification of Occupations, 1960* ("professional" and "intermediate non-manual"); working class defined as groups 6 and 7 ("foremen and skilled manual" and "semi- and unskilled manual").

different, and thus supported Pahl's (1965b) contention that "it is class, rather than commuting characteristics alone, which is the most important factor in promoting change in the social structure of villages in the rural/urban fringe of metropolitan regions" (p. 8). Contrasts in rates of car ownership between members of the two classes were very important in the sample villages, with the proportion of middle-class families which owned cars being double that of the working class in the early 1960's. Middle-class families were thus afforded considerable spatial mobility and could exercise choice in their social and shopping activities. By contrast, at the time of the survey, few working-class wives could drive, and very often there was only one car to a family. This was normally required by the husband to drive to work. The net result of this situation meant that "in their relations with the outside world the middle class and working class villagers moved in separate worlds. . . . The world of the working class was as enclosed and traditional as in any village in the more remote rural areas" (Pahl, 1965b, p. 11). D. C. Thorns' (1968) study showed a similar contrast, with farmworkers in peri-urban areas being strongly village-orientated. "The village is the centre of their lives. They are very often born within a few miles of where they now live. They are restricted in outlook and their involvement outside the village area." But for the middle-class commuters, "their sphere of associations and contacts is wider than the village community which is seen largely as a dormitory and a place to spend the weekend" (p. 163).

Other case studies suggest that this final point may be rather an extreme interpretation. The results of an analysis by A. R. Emerson and R. Crompton (1968) of the social life of newcomers in East Anglian commuter settlements showed that members of one-third of the households investigated spent most of their spare time in village-based activities. An equal proportion concentrated on activities in the nearest town. The remainder spent most of their spare time on home-orientated activities. Perhaps the important thing to stress is that middle-class, car-owning newcomers had the opportunity to choose between town- or village-based activities.

Less affluent village residents, often without their own means of transport, were often denied such a choice.

Other aspects of social activity showed contrasts in behaviour between members of the two classes. For example, there was frequent visiting between parents and siblings amongst members of the working class, but formal entertaining was rare. By contrast, middle-class families indulged in more formal "gracious living" entertaining through the medium of dinner parties.

In some instances a clear spatial separation existed between the sections of commuter villages that were occupied by middle-class and by working-class families in the form of estates of privately developed housing for managerial and professional people and estates of local-authority housing constructed for the working class. This point was well exemplified in the study of Tewin, where Pahl (1965a) concluded that social and spatial segregation was "one of the most important characteristics of the rural/urban fringe" in Britain (p. 46).

Members of the two classes behaved differently in their support for local clubs and social activities. Many middle-class newcomers had sought to enter into a closely knit village community but by their very presence they were contributing to its destruction. In Pahl's sample villages most local clubs and societies were run and supported by members of the middle class. The exceptions to this were the football and old folks' clubs, which were essentially working class, and the horticultural society which involved members of both groups. Even the Church was predominantly a middle-class institution, although with some appeal for the "respectable, conservative working class" (1965b, p. 16).

However, other researchers have stressed that voluntary associations can play an important part in bringing newcomers and established residents together in expanded rural settlements on city fringes. G. Popplestone (1967) enumerated the potential importance of such associations in the following terms. First, they can function as substitutes for kin groups which most newcomers leave behind when they move into the countryside. Second, they can help individuals to learn new norms which are necessary in the new environment. Third, they may avoid possible conflict between newcomers and established residents as they establish contacts through participation in voluntary associations. Finally, they can help newcomers to become identified with the village and with the organization of village affairs.

R. E. Pahl's seminal work depicted the transformation of the traditional hierarchical social structure into a polarized, two-class community, more typical of urban than of rural conditions. But later studies have shown that this process can take on more complicated forms in areas with greater cultural diversity than the environs of London. G. J. Lewis (1967), for example, has shown how complex the "them" and "us" dichotomy can be in settlements in central Wales within the commuting hinterland of a larger urban centre. The basic division between newcomers and established citizens was underlain not only by variations in class and degree of participation in local associations, but also by differences in language and religious observance. Most newcomers were not Welsh-speaking; however those who did speak Welsh were more readily accepted into the existing (predominantly Welsh-speaking) social organizations. Attendance at religious services in commuter villages was lower than would have been expected in traditional Welsh communities. An individual's linguistic competence would help condition the ease with which he was accepted into the local community, but his class would affect where he lived in the village and the amount of power he would receive in the chapel and in various local organizations.

Important variations in the impact of newcomers in commuter settlements have been shown by D. C. Thorns (1968) to depend on variations in the local social structure. Four components were identified in the population of eleven villages close to Nottingham: professional and managerial workers and their families; other white-collar workers; skilled and semi-skilled manual workers; and farm-workers and unskilled workers. The fourth group was decreasing in both absolute and relative terms. The arrival of professional and managerial commuter families changed the behaviour patterns of the established residents. The established middle-class residents identified more closely with the middle-class newcomers, thereby emphasizing differences with other residents. In addition to stressing differences *between* classes, there were also important differences emerging *within* classes, e.g. between skilled and semi-skilled manual workers, on the one hand, and farmworkers, on the other. The second group was relatively disadvantaged in terms of income, housing, hours of work, and conditions of employment by comparison with other members of the working class.

An invasion by middle-class commuters can produce varying social changes linked to differences in characteristics of the status and leadership structure within the village. Thorns identified three types of settlement. In the first, the status structure had been established over a long period and had a farmer or local landowner at its head. There was little or no change following the arrival of commuters. The *status quo* was accepted by them but was likely to undergo change when the present leader died. In the second type of settlement, the established social structure had broken down recently and the community was in a state of transition since those who had inherited leadership functions had failed to fulfil them effectively and thereby lost the respect of the village population. In the third situation the established social structure had also broken down but new leaders had come forward to fill the vacant roles so that new status heads had emerged. Reworking the terminology proposed by G. D. Mitchell (1950) and discussed on page 20, the first type of village was "closed and integrated"; the transitional village could be either "open" or "closed" but would be "disintegrating"; whereas the third (or re-established) village would be of the "open and integrated" type.

Underlying such profound social changes in commuter settlements is the rather paradoxical reaction of vocal middle-class newcomers in defending the countryside. As Masser and Stroud (1965) observe: "newcomers . . . are frequently the most vociferous in wishing to protect whatever is left of the rural atmosphere" (p. 111). Such folk are keen to preserve and to protect as they realize their desire to live in a village with its neighbourliness and special feeling of identity. But respondents to an investigation undertaken by the National Council of Social Service (1963) regretted that "It is not always the same people, however, who lead others in trying to solve problems which affect the original village dwellers very deeply. One example is the problem of transport in rural areas accentuated by the closing of many rural railway lines. This does not affect those who own cars, who are generally also the people who live in the country and work in the towns. But it does affect the old and those who cannot afford their own means of transport" (p. 32). The implications of this recent decline in rural transportation will be discussed later in Chapter 12.

In spite of local variations in social conditions, one must agree with R. E. Pahl (1965a) that "the outer part of a metropolitan region . . . is a frontier of social change, moving over communities and creating as it were new places, which in turn form the bases for different types of community" (p. 73). But urban influences can be diffused into the countryside in the reverse fashion as farmers take on urban/industrial employment but continue to work their holdings.

WORKER-PEASANTS

The Formation of the Worker-peasantry

In addition to the dispersion of the city into the countryside which results from city dwellers moving into villages and becoming commuters, one must consider the mental urbanization of members of the agricultural population who commute to urban industrial jobs each day and yet continue to work their farms in the evenings, over weekends, and during annual holidays from the factory. Such agriculturalists have become "worker-peasants". C. Mignon (1971) has fitted the worker-peasants into a threefold classification of part-time farming in Europe. In the pre-industrial period farming was often complemented by non-agricultural employment in craft activities during slack periods in the agricultural year. In the nineteenth and twentieth centuries worker-peasants have combined full-time factory work with farming. The third type of part-time farming has developed only in recent years and involves the operation of hobby farms. This is just one component in the complicated process whereby city dwellers are rediscovering the countryside.

Part-time farms are not distinguished from full-time holdings in most national collections of agricultural statistics. For this reason worker-peasants may not be recognized from official statistical sources and have not, therefore, received the systematic study which they deserve. Nevertheless, worker-peasants are found in many parts of Continental Europe either where industrial development or mining activity has been implanted in the countryside or else where public and/or private transportation is sufficient to permit long-distance commuting from rural areas to city-based factories.

Worker-peasant holdings certainly existed immediately around coal-mining areas in France, Germany, Poland, and other European countries during the late nineteenth century. Part-time farming, with the participation of the head of the family and often other members in non-agricultural pursuits, was known in many other parts of Europe before World War II. But the real rise of this kind of binary economy dates from post-1945 industrialization. Not surprisingly, concentrations of worker-peasant holdings are found around newly industrialized areas. Public and private transportation facilities have improved through time and thus hinterlands containing worker-peasant farms have become increasingly large.

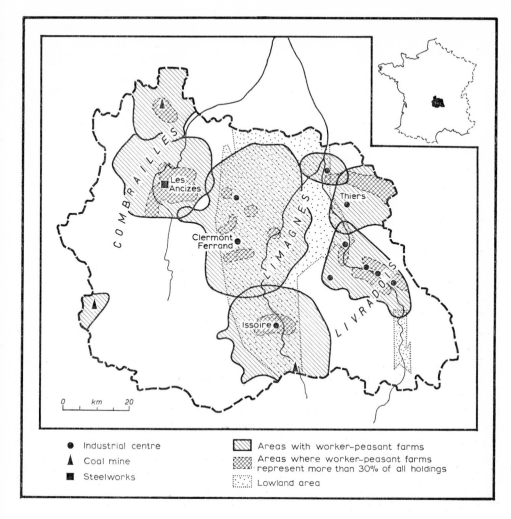

Fig. 4.3. Worker-peasant farming in the Puy-de-Dôme *département* of the Massif Central.

In the nineteenth century they were restricted to walking distances between factory or mine and farm. Now one-way distances of 25 km or over are not uncommon.

In the early 1960's one-quarter of all "farms" in West Germany were worker-peasant holdings, and in Poland the figure was as high as one-third. In some parts of the French Massif Central about one-quarter of all holdings were estimated to be worker-peasant farms. Figure 4.3 shows their distribution in the *département* of the Puy-de-Dôme where worker-peasants commute to the great rubberworks of Clermont-Ferrand, the metalworks of Issoire, steelworks at Les Ancizes, small coal-mines, and industrial centres elsewhere in the area. Up to one-quarter of all holdings being operated by worker-peasants were also found in the Plain of Alsace, central Lorraine, and the industrialized valleys of the northern Alps. In the German Saarland almost one-half of all farms were worked by worker-peasants.

Rural residents have chosen to remain living on the land, and yet commute long distances to work each day for numerous reasons. These include sentimental attachment to their family land as well as the availability of adequate transport which puts factory wage rates within the reach of the countryman. In Germany, for example, rural residents had endured severe monetary devaluation, destruction of property during World War II, and also serious shortages of food. Nevertheless, the countryside still offered very real attractions. Land tended to retain its value and was a far surer investment than depositing in city banks. Also one had the opportunity of growing one's own food from one's own land or, at second best, of being close to the farmers who did produce food. The countryman therefore ran far less risk of going hungry than did his city cousin. Many rural residents were positively attracted to remain on the land and postpone ideas of migrating to the city. R. Klöpper (1971) reports that during the inter-war period dwarf farms were not abandoned in Germany. On the contrary, their cultivation was sponsored by the authorities as a supplementary means of existence for those threatened with unemployment or with the prospect of becoming homeless. Many people learned that both security and adequate living conditions could still be found in the countryside. Such attractions were particularly strong after the war for refugees from eastern areas for whom the search for security was particularly dear.

Almost all the larger German towns had suffered severe bombing during World War II. Serious housing shortages resulted in urban areas. Commuting from the countryside was sponsored by the authorities as a way of reducing the accommodation problem and yet obtaining the necessary supplies of labour for industrial revival. Indeed, the recovery of factory production in Germany and Poland and the beginning of industrialization in Yugoslavia and other parts of eastern Europe was only achieved by using large amounts of worker-peasant labour as small farmers travelled considerable distances to factory work each day. For example, in 1957 the major Yugoslav industrial centres of Belgrade, Ljubljana, and Zagreb drew more than 10 per cent of their daily migrants from settlements more than 50 km distant (Franklin, 1969). It is now likely that the expansion of urban housing programmes has changed this situation and permitted a reduction in the numbers of long-distance worker-peasant commuters.

Advantages and Disadvantages of Worker-peasant Farming

Both advantages and disadvantages stem from worker-peasant farming. It might be argued that the existence of such a dual economy provides a "cushion" in times of industrial hardship and unemployment (as was the case in Germany) so that the worker-peasant may return to working the land for a while. Worker-peasants undoubtedly gain higher incomes than could be derived either from just farming or from industrial work. This extra income might be used to improve the family's living conditions or the equipment on the farm or might be spent on the children's education. The existence of worker-peasant communities in the countryside means that the demand for local services will be kept up and hence rural shops, schools, and other facilities will be retained to benefit agricultural workers and other rural dwellers as well as the worker-peasants and their families. Ruralization will be avoided. It might also be argued that the worker-peasants' new contacts, derived from their participation in industrial as well as agricultural work, would provide a richly stimulating social environment to diffuse new ideas into the countryside for the improvement of farming, education, and social conditions.

These possible advantages are normally outweighed by the numerous disadvantages of the worker-peasants' way of life. Their holdings vary considerably in size, but it is clear that labour inputs are restricted and, hence, only small units may be managed. Such holdings are often fragmented into a number of small blocks or strips. This kind of property fragmentation on part-time farms hinders strip consolidation and farm-enlargement schemes which would benefit full-time farmers.

Worker-peasant farms are often over-mechanized. The "5-o'clock farmers" are relatively well off with two sources of income and can therefore afford to purchase tractors more readily than many full-time farmers. Tractors and other pieces of machinery are viewed as status symbols for acquisition. In fact they may be used only a few times a year on the small part-time farms and are thus uneconomic.

Farms operated by worker-peasants are often of low productivity because their operators have had little or no formal agricultural training and their labour inputs are severely restricted. In Poland worker-peasant production concentrates on less-demanding crops such as rye and potatoes (Zajchowska, 1971). C. Mignon, working in the French Massif Central, has shown that the agricultural systems practised by worker-peasants varied considerably between localities in the region, but they were characterized by the common feature of being less intensive than those practised by full-time farmers. Cereal yields and livestock densities on worker-peasant farms were well below the regional average. Worker-peasant farms that raised livestock tended to rely on permanent grass as the sole source of fodder rather than making use of more sophisticated sources such as ley or rotated grass.

Worker-peasant farms, wherever they are located, are normally much smaller than the average for the region concerned. Even so, sections of farmland may fall out of agricultrual use since insufficient time and labour can be devoted to their cultivation. J. Labasse (1961 and 1966) describes the type of situation whereby the old traditions of full-time farming gradually collapse and are replaced by extensive grazing or the small-scale production of vegetables. Even this may then give way to almost complete land abandonment with only the kitchen garden being used. This type of phenomenon is known as social fallow.* Property is left uncultivated not because of any inherent poverty of the soil but because of social reasons already mentioned. Up to 50 per cent of the agricultural land surface has been abandoned in some German villages which contain large numbers of worker-peasants. In parts of the Massif Central, where part-time farming is practised, such under- or unused land will degenerate to rough scrubby pasture or, alternatively, may be

* *Jachère sociale* in French; *Sozialbrache* in German.

planted with timber. In any case, its existence will hinder schemes for farm enlargement, which would benefit full-time farmers, and will make agricultural restructuring expensive to implement. Owners of surrounding properties, which may still be in agricultural use, will inevitably suffer as weeds and pests invade their land from patches of social fallow held by worker-peasants. As S. H. Franklin (1969) has rightly noted, "the development of a worker-peasant class solves the problems of the agriculturalist rather than agriculture" (p. 212).

Worker-peasants are neither complete farmers nor complete industrial workers. Because of their "mixed" status they do not qualify to receive the types of social benefit paid to members of either group. Worker-peasants' lives are very hard. As industrial workers they will be committed to their 8-hour shift work for 5 or 6 days each week. On top of this many of them will have to spend up to 2 hours commuting between farm and factory each day. Free time has to be devoted to running the farm in the evenings and over weekends. Not surprisingly the proportion of worker-peasants in the Massif Central who are unmarried is high. Few young women are willing to choose a life of "veritable slavery" with little or no leisure time, as has been described by C. Mignon. In fact, many of the worker-peasants he studied abandoned part-time farming when they married and became either factory workers who commuted from the countryside but no longer worked the land, or else full-time city dwellers.

It might be thought that the coexistence of agricultural and industrial ways of life in a dual economy would allow worker-peasants to enter a broad social environment and thereby gain new urban ideas to allow them to improve their rural existence. Franklin (1969) has summarized what happens. By entering the factory doors the farmer "opens himself and his family to the influence of foreign attitudes, standards and practices, each possessing a certain cachet in the modern world against which traditional attitudes towards work and its returns must inevitably be compared. How long it will take them to erode old attitudes no one knows, but erode they must and with that erosion will come the abandonment of farming, and later perhaps the disposal of the property" (p. 55).

Nevertheless, the erosion process is often a protracted one. Worker-peasants are divided in their allegiances. They retain territorial ties with the land they own and work and they also maintain family ties and religious observances in their home villages. In addition, new occupational contacts are gradually formed in the factory and in the town to which they commute each day. Thus there is not a sudden break between rural and urban outlooks and ways of life. "The worker-peasant class remains part of the socio-economic life of the rural and farming community, and only slowly through its different experiences and the divergent attitudes which result does it become an agent of change" (Franklin, 1969, p. 56). Many worker-peasants are politically docile and socially unmixable. They rarely belong to industrial trades unions and are equally unwilling to enter agricultural co-operatives. This can act as a major hindrance to agricultural improvement in areas where large numbers of worker-peasant holdings are found. Worker-peasants are, however, very keen on investing in their children's education so that they can escape from the insecurity and split-personality existence of being neither a true farmer nor complete industrial worker.

The widespread provision of public transport services and works buses permitted long-distance commuting and the transformation of full-time farms into worker-peasant holdings in areas far from central factories. In western Europe this process was aided by rising rates of private-car ownership. To take a single example from around the Ancizes steelworks (Fig. 4.3) in the Massif Central, over 80 per cent of farms in communes adjacent to the steelworks are held by worker-peasants. The proportion falls to 50 per cent within a 10 km radius and then gradually declines with increasing distance.

After a period of increasing importance from 1950 to the early 1960's, the worker-peasant phenomenon now appears to be contracting in Continental Europe. In the last decade, worker-peasant holdings have decreased in number. Young people are rarely willing to become part-time farmers for three main reasons. First, dual employment puts the worker-peasant in an insecure position and allows him and his family very little leisure. He owes complete allegiance to neither way of life and is often regarded as inferior both as a farmer and as a factory worker. Second, the children of worker-peasants are now better trained for performing industrial or tertiary jobs than were their parents. Some young people choose to live in the countryside after they have taken on full-time city employment. Mignon reports that if they decide to take on spare-time work this will involve painting, car-repairing, or similar jobs. It will very rarely involve their becoming part-time farmers. Third, sentimental attachment to farmland which may have been held by the family for decades becomes weaker with each successive generation. Modern young people are far less concerned about living on the ancestral farm than their parents were a score or more years ago.

Worker-peasant farming simply represents one component in the complex process of urbanization. Worker-peasant holdings have already disappeared from many immediately peri-urban areas which have been converted into suburbia. Young men are not following their fathers into the binary economy. Most worker-peasants in the Massif Central, for example, are between 40 and 50 years of age. This corresponds with the critical 20–30-year age group in the early 1950's when young men were returning from military service, marrying, and deciding on their future careers in the post-war industrialization period. Few worker-peasants in the Massif are now under 30 years of age. Nevertheless, the mean age of worker-peasants is below that of full-time farmers. "It is significant that worker-peasant families tend to be headed by younger men, which no doubt enables them to bear the strain of a twofold occupation" (Franklin, 1969, p. 211).

The binary economy of the worker-peasant is open to attack from several directions. Many of the manufacturing and mining processes which demanded large quantities of labour in the past and encouraged the development of the worker-peasantry are now shedding employees as mechanization and automation become more widespread. Franklin (1964, 1969) provides the detailed example of the German village of Gosheim on the edge of the Swabian Jura whose workshops producing nuts, bolts, screws, etc., flourished during World War I, the period of Nazi rearmament, and in the post-war period of the economic miracle and the rise of the consumer society. The key to economic success in the past had been large quantities of cheap, worker-peasant labour. Recently, automation has started to replace manual labour. Small firms are being taken over and rationalized by larger ones. The net result is to put members of the worker-peasantry out of industrial employment. A similar problem threatens in steel-making, coal-mining, and many other industrial branches throughout Europe where mechanization and rationalization are destroying the *raison d'être* of the worker-peasantry. Unfortunately, a return to full-time farming will not provide an adequate alternative income for the worker-peasant who has lost his factory job.

Unless new forms of employment can be brought into worker-peasant communities it is likely that many worker-peasants will concentrate fully on newly obtained city jobs after having abandoned part-time farming. They may continue to live in their farmhouse and be long-distance commuters or they may move to accommodation in town. No matter which solution is found, many worker-peasants are unwilling to sell off either all or part of their holdings which may become completely abandoned as social fallow. Many are keen to hold on to their land which may provide future building plots for the former worker-peasants and for members of their families. If the farmhouse is not occupied full time by the owner, leased out to neighbours, or abandoned completely, it may be used as a leisure house or second home over weekends and during vacations. The surrounding land may even be run as a hobby farm.

In fact industrialists are often unwilling to open new factories to take on former worker-peasant labour because of the slowness of such folk in adapting to really modern industrial work and the fact that the worker-peasant system has in the past been characterized by low productivity rates and high absenteeism. Factory work, whether provided at isolated points in the countryside or concentrated in large cities, seems to require a more complete transformation of agriculturalists into industrial workers than their continuance in the dual role of worker-peasants can guarantee (Labasse, 1966).

REFERENCES AND FURTHER READING

A survey of the functional morphology of the rural population in England and Wales in 1951 is contained in:

ROBERTSON, I. M. L. (1961) The occupational structure and distribution of rural population in England and Wales, *Scottish Geographical Magazine* **77,** 165–79.

The now classic analysis of commuter villages north of London is found in:

PAHL, R. E. (1965a) *Urbs in Rure*, London School of Economics Geographical Papers, **2.**
PAHL, R. E. (1965b) Class and community in English commuter villages, *Sociologia Ruralis* **5,** 5–23.
PAHL, R. E. (1966) The rural/urban continuum, *Sociologia Ruralis* **6,** 299–329.

Other regional examples are provided by:

CONNELL, J. H. (1971) Green belt country, *New Society*, 25 Feb. 1971.
CRICHTON, R. M. (1964) *Commuters' Village*, Museum of Rural Life, Reading.
GREEN, P. (1964) Drymen: village growth and community problems, *Sociologia Ruralis* **4,** 52–62.
LEWIS, G. J. (1967) Commuting and the village in mid-Wales, *Geography* **52,** 294–304.
LEWIS, G. J. (1970) A Welsh rural community in transition: a case study in mid-Wales, *Sociologia Ruralis* **10,** 143–61.
MASSER, F. I. and STROUD, D. C. (1965) The metropolitan village, *Town Planning Review* **36,** 111–24.

POPPLESTONE, G. (1967) Planning for the changing countryside. Mimeographed paper for the Town Planning Institute research conference.

Sociological case studies include:

EMERSON, A. E. and CROMPTON, R. (1968) Report to the Suffolk Rural Community Council, Suffolk: Some social trends. Mimeographed report.

THORNS, D. C. (1968) The changing system of social stratification, *Sociologia Ruralis* **8,** 161–78.

A morphological analysis of commuter settlements around London is contained in:

WHITEHAND, J. W. R. (1967) The settlement morphology of London's cocktail belt, *Tijdschrif voor Economische en Sociale Geografie* **58,** 20–27.

Other considerations of social change in the countryside are presented in:

ANON. (1963) *Communities and Social Change: introductory survey*, National Council of Social Service, London.

ASHBY, A. W. (1939) The effects of urban growth in the countryside, *Sociological Review* **31,** 345–69.

BRACEY, H. E. (1963) Rural settlement in Great Britain, *Sociologia Ruralis* **3,** 69–77.

GLASS, R. (1962) Urban sociology, in WELFORD, A. T. *et al.* (eds.), *Society: problems and methods of study*, London.

JOHNSTON, R. J. (1965) Components of rural population change, *Town Planning Review* **36,** 279–93.

KENDALL, D. (1963) Portrait of a disappearing English village, *Sociologia Ruralis* **3,** 157–65.

KURTZ, R. A. and EICHER, J. B. (1958) Fringe and suburb: a confusion of concepts, *Social Forces* **37,** 32–37.

PAHL, R. E. (1970) *Patterns of Urban Life*, Longmans, London.

WIBBERLEY, G. P. (1960) Changes in the structure and functions of the rural community, *Sociologia Ruralis* **1,** 118–27.

A definition of urban commuter hinterlands in Great Britain is found in:

LAWTON, R. (1968) The journey to work in Britain: some trends and problems, *Regional Studies* **2,** 27–40.

The worker-peasant phenomenon in several countries is discussed by:

FRANKLIN, S. H. (1969) *The European Peasantry*, Methuen, London.

Detailed case studies are found in:

CLOUT, H. D. (1972) Part-time farming in the Puy-de-Dôme département, *Geographical Review* **62,** 271–3.

FRANKLIN, S. H. (1964) Gosheim, Baden-Württemberg: a Mercedes Dorf, *Pacific Viewpoint* **5,** 127–58.

JUILLARD, E. (1961) L'Urbanisation des campagnes en Europe occidentale, *Études Rurales* **1,** 18–33.

JUILLARD, E. (1962) L'Ouvrier-paysan en Lorraine mosellane (review), *Études Rurales* **6,** 202–4.

KLÖPPER, R. (1971) The urbanization of rural districts in Western Germany: with special reference to Rheinland-Pfalz, in DUSSART, F. (ed.), *L'Habitat et les Paysages Ruraux de l'Europe*, Liège, pp. 283–91.

MIGNON, C. (1971) L'Agriculture à temps partiel dans le département du Puy-de-Dôme, *Revue d'Auvergne* **85,** 1–41.

ZAJCHOWSKA, S. (1971) Processus d'urbanisation de la campagne en Posnanie, in DUSSART, F. (ed.), *op cit.*, pp. 409–22.

The development of social fallow in Federal Germany is described in:

LABASSE, J. (1961) Structures et paysages nouveaux en Allemagne du Sud, *Revue de Géographie de Lyon* **36,** 93–116.

LABASSE, J. (1966) *L'Organisation de l'Espace: éléments de géographie volontaire*, Hermann, Paris.

CHAPTER 5

URBANIZATION
OF THE COUNTRYSIDE—II

LEISURE AND THE COUNTRYSIDE

Leisure activity is by no means a new phenomenon in the countryside. Traditional sports of hunting, shooting, and fishing have been part of the rural scene for centuries. But the numbers of people involved in such activities were small and, for the most part, there was little widespread impact on rural resources. The recent incursion of urban dwellers has resulted in activities quite different in kind and scale, and hence there has been a correspondingly much greater impact on the countryside than ever before (Hookway and Davidson, 1970). In the late nineteenth century the railway and the steamship permitted movement on a hitherto unprecedented scale. Also the rise of an affluent stratum of society between the landed aristocracy and the mass of the people provided an ever-growing supply of urbanites in search of recreation. On the eve of World War I, France, Italy, and Switzerland already derived substantial revenues from middle- and upper-class holiday-making. After wartime disruption, tourism recovered with an inter-war peak being reached in 1929 on the eve of the great depression. Since World War II the situation has changed dramatically as recreation has been transformed into a mass activity indulged in by most groups in society. As a result, ever-increasing numbers of visitors spend their weekends and vacations in the countryside as well as on the coast or at mountain resorts (Saville, 1966).

Such an explosion in numbers has resulted from four factors:

(a) rising living standards which have accompanied high rates of economic growth in Europe and North America;
(b) the widespread extension of holidays with pay;
(c) the continuing expansion of mass-travel facilities, especially the private car; and
(d) the large-scale organization of recreation facilities.

Not until 1936 did the International Labour Office adopt its draft convention for holidays with pay. At that time the six countries of the (present) Common Market contained no more than 300,000 workers with paid leave as part of their labour contract. Over 1 million people had such contracts in the United Kingdom in the mid 1930's. Nearly all manual workers in western Europe now have at least a fortnight's paid leave and in many countries the trend is towards a 3 or

4 weeks' minimum, as is already the case in France and Sweden. Variations in the mean duration of paid leave clearly condition the length of main holiday spent away from home. In 1960 this varied from 22 days in France (Table 9) to only 10 days in the Netherlands and the United Kingdom. But at the same time there were important national differences in the proportion of the population going away on holiday, ranging from 21 per cent in Italy to 60 per cent in the United Kingdom and 72 per cent in Sweden (Table 10). Normally there is a direct relationship between the proportion of the population living in cities and the volume of people going away on holiday.

TABLE 9. MEAN LENGTH OF MAIN HOLIDAY,* c. 1960

	Days		Days
Belgium	15	Netherlands	10
France	22	Sweden	12
West Germany	19	United Kingdom	10
Italy	11	United States	11

* Defined as more than four consecutive days away from home.

Source: CRIBIER, F. (1969), *La Grande Migration d'Été des Citadins en France*, CNRS, Paris, p. 57.

TABLE 10. PROPORTION OF NATIONAL POPULATION GOING
AWAY ON HOLIDAY,* 1960–5

	(%)		(%)
Belgium	48	Netherlands	55
France	45	Sweden	72
West Germany	37	United Kingdom	60
Italy	21	United States	43

* Defined as more than four consecutive days away from home.

Source: CRIBIER, F. (1969) *La Grande Migration d'Été des Citadins en France*, CNRS, Paris, p. 56.

In addition, increasing lengths of time are being spent in the countryside by city dwellers over weekends and during public holidays. The latter vary in frequency according to national variations in religion and tradition, reaching, for example, 13 days per annum in West Germany and 17 days in Italy. However, there has also been a marked trend for the length of the working week (here defined as the average weekly numbers of hours in manufacturing industries) to fall throughout western Europe. The following figures indicate changes between 1960 and 1967: West Germany, 45.6 to 42.0 hours; Netherlands, 48.8 to 45.3

hours; and the United Kingdom, 47.4 to 45.3 hours. At the same time, rates of car ownership have risen dramatically. Thus the number of cars in sixteen west European countries was 21.5 millions in 1960 (Fig. 5.1). Seven years later it had increased to 50.3 millions (+133 per cent). Larger numbers of increasingly mobile city dwellers are spending their weekends and leave-days in the country-side. Heavy pressures on recreational resources build up at viewpoints, historic and architectural monuments, and woodland and water areas within easy access of cities. But increasing numbers of city people are also moving for weekend recreation into areas which lack these attractions.

Catering for visitors gives rise to important social and economic changes in localities where it has become established. The following discussion will concen-trate on the types of change that have been monitored in Alpine areas where catering for winter-sports enthusiasts and for summer visitors has introduced important features of innovation and has diffused urban ideas. Most of these changes have a more general applicability than in that particular research area. J. Saville (1966) has summarized the kinds of change that take place: "In general terms it is difficult to exaggerate the effects of mass tourism in the next 100 years upon the living pattern of rural society over most of Europe. For peasant societies, in particular, the impact will be a powerful one. . . . Tourism can only increase the urban encroachment upon the countryside; it will greatly encourage the pro-cess of mental urbanization; and it will be a powerful factor in the transformation of 20th century rural society into social forms and structures which, by the mid-21st century, will be radically different from those which obtain at present. It will, in short, powerfully assist the accelerating trends towards the urbanization of the countryside" (p. 34).

The presence of visitors in a rural community for part of the year leads to a revival of local life as visitors stay in hotels or boarding houses or live on camp sites or in second homes. Very often the presence of city people provokes changes in the appearance of the invaded settlements as buildings are repainted, old houses renovated, and new ones constructed. But catering for visitors is perhaps even more important for the intangible social and economic changes it stimulates as urban capital and ideas are injected into the countryside.

The rural population is normally unprepared to welcome tourists, and there may well be an initial period of hostility towards the "outsiders" who are treated quite differently from members of the village community (or their children) who have migrated away in search of urban employment but who come back to the home community for their vacations. Such people are often welcomed by the established village residents as lost sheep returning, if only temporarily, to the fold. The initial hostility expressed towards outsiders will diminish as members of rural communities realize that catering for them can provide at least four benefits.

First, it gives rise to supplementary forms of employment which may be especi-ally beneficial to women and to young people who always form categories for whom jobs are not readily available in farming-based areas. Thus P. Rambaud (1969) reported that country dwellers considered that the creation of additional forms of employment was the most important advantage derived from the presence of visitors in the countryside (Table 11). Work is available in hotels and boarding

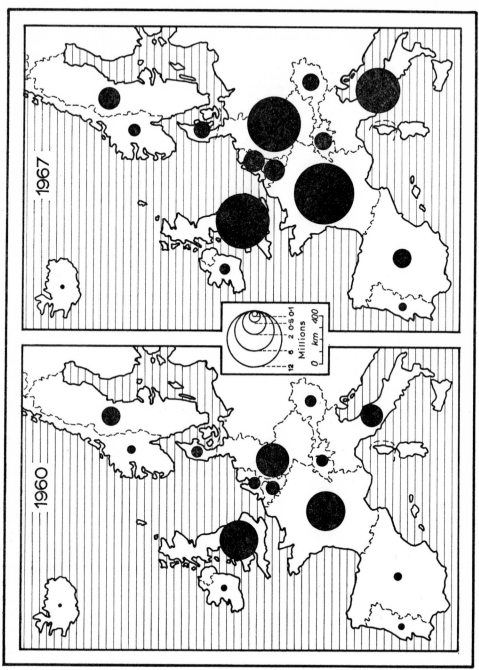

Fig. 5.1. Numbers of cars in western European countries, 1960 and 1967.

TABLE 11. ADVANTAGES DERIVED
FROM PRESENCE OF VISITORS IN THE
COUNTRYSIDE

	%*
Creation of new jobs	50.0
Provision of extra income	25.0
Introduction of new ideas	22.5
More social life	10.0
No opinion	2.5

* Total exceeds 100 per cent since some
replies were joint answers.
Source: RAMBAUD, P. (1969), *Société rurale
et urbanisation*, Seuil, Paris, p. 160.

houses or in the production and sale of souvenirs, such as ironwork, pottery, handwoven cloth, etc.

Second, these jobs can provide important additional incomes for rural families. For example, rooms may be let in local farms and other houses to accommodate visitors. As J. Morgan-Jones (1972) has shown, "a farm tourist industry calls for an entirely new attitude towards the use of time, of premises, and of the farm house all of which, of course, is of vital concern to the farmer's wife and daughter" (p. 122). Michael Dower (1972) suggests that a high proportion of the total expenditure by visitors is paid for labour and for local products and services rather than in petrol taxes and other forms which go straight out of a tourist region. He estimates that of the total expenditure by a vacationist, once he has reached his holiday region, "probably 50–60 per cent is spent on accommodation and food, about 15 per cent on travel within the region, and the rest on recreation, shopping, entertainment, eating out, etc." (p. 81).

Third, local farmers have the chance of disposing of some of their produce (milk, vegetables, fruit) to local hotels, the occupants of second homes and itinerant visitors, thereby obtaining additional revenue. Finally, contacts between visitors and members of rural communities can be beneficial, bringing "a wider experience and understanding of life and work in the world outside" (Bracey, 1970, p. 257).

The rural dweller's conception of "work" will be changed after he has made contact with urban visitors. In the past, traditional agricultural activities were viewed as a way of life rather than a business activity. As such, farmwork was organized to occupy the farmer's waking hours with varying degrees of intensity throughout the day and throughout the farming year. "Work" was not separated from "leisure" as is the case in industrial and post-industrial society with a 5-day, 40-hour "working week" (or whatever) being separated from one's "free time". But once farmers and members of farming families start catering for visitors they will reorganize their activities towards overt profit-making and also so that there is a distinct break between "work" and "leisure".

Farming will be thrown into a state of inferiority since catering for tourists has four major features in its favour. First, it is normally more financially reward-ing than working the land. Second, it involves a degree of professional training which farming does not, for example, as a farmer works as a ski-instructor or as a waiter in the tourist season. Third, meeting the needs of visitors confers a degree of social prestige on the individual and gives him the chance of social advancement since his skills might be used more intensively (and more profitably) either in his home community or in a larger settlement, whether this be another tourist resort or a city. Finally, catering for visitors permits a profit to be made with fewer uncertainties than would be encountered in farming.

Thus tourism, like all forms of urban influence in the countryside, works to break up elements of cohesiveness in village society by introducing or intensifying competitiveness and individualism. Where rural residents may have co-operated in farmwork in the past they will now act as rivals as they compete to try to satisfy visitors' demands. But not all members of rural communities wish, or are able, to compete in this respect, and thus jealousy may emanate from those who have not benefited towards the farmers, shopkeepers, and small hoteliers who have improved their income and changed their way of life.

The rural environment of farms and fields will be viewed in markedly different ways by farmers and by tourists. To the visitor a stretch of farmland may be visually attractive or unattractive. It may or may not offer facilities for his enjoyment of the area. By contrast, the same stretch of countryside represents the environment from which the farmer has to make a living. Rural areas are thus evaluated differently by members of different groups who have varying objectives in mind. Such areas are also perceived in very different ways. The farmer views a section of farmland as a compartmentalized stretch of countryside, divided into fields and plots of land with clear differences in fertility, traditional function, and sentimental associations. He is thus adopting an ecological approach to the area in question, realizing how different fields are used for varying purposes throughout the year on his own farm and how his fields are arranged in relation to the pro-perties of his neighbours. The visitor tends to view the rural environment in quite a different way, seeing it as a series of isolated points in space which might be developed for his enjoyment. He is not concerned with, or indeed often aware of, the interrelations with surrounding plots of land. Two completely different sys-tems of values are thus being applied to sections of the rural environment. For example, a farmer may perceive a steep hillside as an infertile tract of land with thin stony soils which are of limited value for farming purposes. On the basis of such criteria, he will award it a low monetary value. On the other hand, an urban visitor may view the same stretch of land as a highly desirable site with an excellent view over pleasant scenery and offering a good location for constructing a second home. As a result, visitors are willing to offer rural landowners far greater sums for the sale of land than could be obtained if the property were being simply sold off for agricultural use.

A range of new relationships develop once a rural area becomes part of urban leisure space. The village and its inhabitants are no longer concerned simply with the production of agricultural goods but have entered into the national, and, indeed, international recreation environment. Employment patterns are changed as supplementary jobs are introduced and new periods of intensity are created. The rural environment, which was perhaps marginal for farming purposes and appeared "repellent" to the local population (who responded by outmigration), has now become "attractive" to visitors and can be made to appear even more so by the provision of additional facilities by rural residents. As a result, fewer decisions in village life are taken on the basis of traditional, community-centred criteria. Instead, more and more are geared to meeting the tastes and desires of urban visitors.

Conflicts will inevitably arise since not all village residents wish to sell their service to visitors or dispose of their property for recreational use. Such folk may find that their schemes for farm enlargement or plot consolidation are frustrated, first, by the scatter of land around the village owned by urban visitors and thus not available for agricultural restructuring, and, second, by the general rise of land values in the village once the urban invasion has started to take place.

The net results of the social changes resulting from the impact of visitors on rural communities are open to debate. On the one hand, one can argue that catering for visitors provides a range of employment to complement farming and thereby offers additional sources of income which will allow rural dwellers to improve their living conditions and thereby encourage them to remain in the countryside. Such an advantage might be of particular value with respect to young women for whom local jobs are not easily available and who are thus more prone to migrate to the city than are young men. On the other hand, it could be argued that the diffusion of information on urban ways of life through contacts with visitors will only serve to make country dwellers less content with their lot and encourage them to move to the town where their newly acquired skills may be perfected and put to even more profitable use. This might be particularly true of young people who are far less tied to their home environment than were their parents or grandparents. In fact, one can produce examples to support either point of view.

One thing is certain: "The life of the village and its social psychology will be more or less altered by the influence of the outside world and by the new orientation of the economic life of its rural community. The reshaping of village life may be only partial or it may in the end be complete. The thoroughness of the change will largely depend on the degree to which the traditional economic pursuits of the villagers continue or are superseded by the demands of the tourist trade" (Saville, 1966, p. 32). Some parts of Alpine Europe, for example, contain settlements with two tourist seasons—summer and winter—where traditional pastoral activities are no longer necessary to most farming families, but they may well be continued for other, nostalgic reasons.

Successful holiday visits to rural areas may lead to the establishment of more permanent connections between urbanites and the village in question. Middle-aged visitors may decide that they will retire there in their old age. Such migrations of elderly people create a local demand for service provision and may well maintain some rural dwellers in the countryside who might otherwise have left. Alternatively, short-stay visitors to the rural area may decide to purchase a cottage or have a new house built as a second home for use on a regular basis during vacations and perhaps also over weekends. This type of urbanization of the countryside forms the final theme of this section.

SECOND HOMES

Seasonal Suburbanization

The establishment of increasing numbers of second homes for weekend and vacation use in rural areas in developed parts of the world may be viewed as a variant on two aspects of urbanization. On the one hand, it may be seen as a specialized form of holiday accommodation for visitors. This interpretation would apply particularly to areas remote from large urban centres and thus beyond hinterlands for weekend recreation. On the other hand, the proliferation of second homes may be interpreted as a seasonal or temporary extension of suburbia, as cottages are occupied for 2 or even 3 days each week virtually throughout the year. In some regions of the developed world, such as parts of North America and North-west Europe, "going up to the cottage" is accepted by many families as part of their normal residential activity. It is not equated with "going away on vacation". Particular emphasis will be placed on this second type of interpretation in the discussion which follows.

A wide variety of types of dwelling might be covered under the umbrella term of "second home", including caravans and even houseboats. The following consideration will disregard such mobile dwellings. Even so, a wide range of buildings has to be included, from converted cottages formerly occupied by fishermen and farmers through do-it-yourself second homes, to architect-designed villas and chalets. The distribution of second homes in the countryside varies enormously from place to place. Second homes may cluster together or be widely dispersed in the open country. They may form parts of existing settlements, made up predominantly of primary residences, or they may be set apart on specially built estates. They may be carefully designed to blend into rural landscapes or they may stand out as unwelcome intrusions.

The popularization and proliferation of second homes is essentially a post-1945 phenomenon, resulting from a combination of sufficient income to be devoted to non-essential items and sufficient free time for this income to be used on leisure activities. Improvements in public and especially private transportation have allowed individuals to realize their own specific motives for second-home acquisition, such as following fashion, wanting to engage in non-urban recreation, or desiring to invest one's savings in property.

In the past, second-home occupation was limited to a very small and affluent section of society. The nobility and bourgeoisie in England, France, Sweden, and other European countries acquired weekend and summer retreats from the seventeenth century onwards. The restricted nature of second-home living has been strikingly changed since World War II as second homes have been purchased by growing numbers of middle-income families. Even so, the more affluent groups in society predominate in second-home living, as Table 12 suggests. This shows the percentage of households by socio-economic groups which owned second homes in

France in 1964. Melvin Webber (1968) has predicted that recent trends for increased personal mobility will intensify so that multi-house and multi-car families will become far more numerous in the future and thereby bind town and country together in large functional hinterlands. Such an ecological interpretation would view settlements in the countryside which contain second homes as highly specialized extensions of the city's living space.

TABLE 12. PROPORTION OF HOUSEHOLDS OWNING
SECOND HOMES IN FRANCE BY SOCIO-ECONOMIC
GROUPING (1964)

	%
Farmers	1.0
Agricultural wage earners	0.0
Employers in industry and commerce	12.1
Liberal professions and top-level management	28.8
Medium-level management	13.8
Clerical workers	10.0
Manual workers	5.2
Service personnel (domestic)	2.0
Other employed persons	11.1
Non-active personnel	4.7

Source: CLOUT, H. D. (1969), Second homes in France, *Journal of the Town Planning Institute* **55,** 440.

Primary suburbanization permitted spatial separation between place of work and place of residence within the framework of the built-up city. Second-home occupation has introduced a new dimension to this process, allowing the urban dweller to change his spatial environment completely on a regular and repeated basis rather than simply moving from one part of the built-up city to another as a result of his journey to work. The temporal distinction between first and second homes is decreasing and will continue to do so as working weeks become shorter in the future. Second homes within easy access of urban demand centres are already being occupied by city dwellers for two and increasingly for three nights each week virtually throughout the year. Second homes in more remote areas may be occupied continuously during long vacations. In many developed countries this may be for two or three months throughout the period when the schools are closed. The mother and children will live in their part-of-the-year residence throughout the summer and be joined by the father over weekends and during his official paid leave. Many second homes are well equipped with domestic appliances to allow city dwellers to enjoy their stay in the countryside (Table 13).

Little serious attention has been paid to the existence of second homes until very recent years. Detailed rates books or land-taxation registers exist in many countries and allow property owned by non-residents to be identified. In many, but not all, instances such houses are used as second homes. Figures 5.2 and 5.3 have been constructed from the information contained in rates books for five

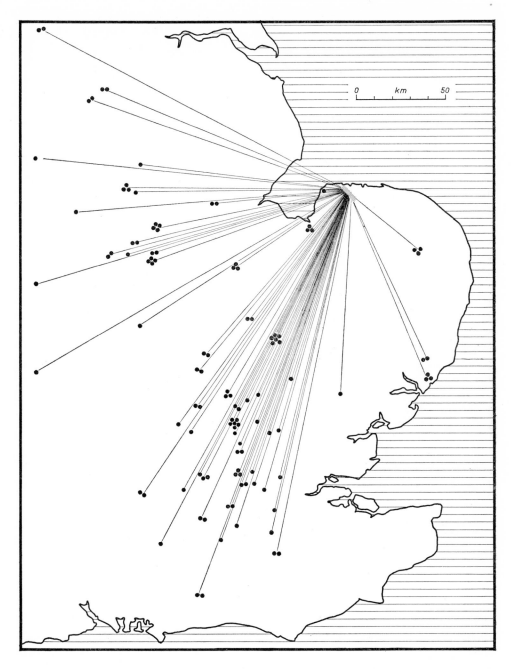

FIG. 5.2. Distribution of primary residences of owners of second homes in five villages in north Norfolk, 1970.

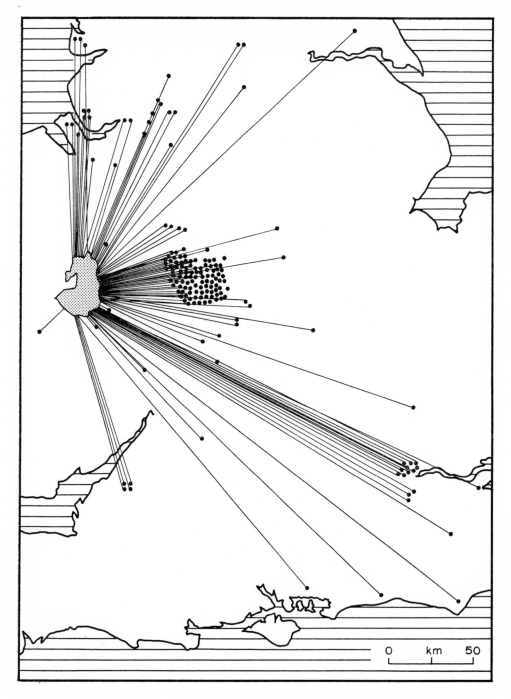

FIG. 5.3. Distribution of primary residences of owners of second homes in the Clun and Bishop's Castle area in Shropshire.

villages on the north Norfolk coast and in the Welsh borderland. They show that relatively remote parts of the country have been selected by Londoners and city dwellers in north-west England and the Midlands for acquiring second homes. Many other parts of the English and Welsh coasts, together with upland areas such as central Wales, the Lake District, and parts of the Pennines, contain second homes. The same is true for country areas within easy access of large urban centres. A detailed study undertaken by the country planning staff in Denbigh-shire describes the situation in the Welsh borderland. Unfortunately, a fully comprehensive survey of second homes in Great Britain has yet to be undertaken. J. Barr estimated that in 1967 only 1 per cent of households possessed second homes; however, one suspects that the proportion is increasing rapidly.

TABLE 13. DOMESTIC EQUIPMENT
IN FRENCH SECOND HOMES (1967)

	%
Piped water inside house	85
Electricity	96
Refrigerator	56
Radio	64
TV set	34

Source: LE ROUX, P. (1968), Les Résidences secondaires des Français en juin
1967, *Études et Conjonctures*, Supplément, **5,**
25.

In other countries national inventories of second homes have been drawn up. In Sweden, for example, 300,000 second homes were enumerated in 1962 with important concentrations around Göteborg, Stockholm, and other major cities (Fig. 5.4). By 1969 the figure had risen to 450,000 when one urban family in every five made use of a second home. In some countries specific questions on second homes have been included in successive censuses. Thus 54,000 second homes were recorded in Belgium in 1961 and 1.23 millions in France in 1968. Official esti-mates have raised this figure to 1.6 millions in 1971 when 20 per cent of French households had access to second homes. This represented a dramatic increase over the figure of 447,000 in 1954. Census data in Australia and in the United States do not relate specifically to second homes which, however, may be identi-fied by manipulating statistics for associated categories listed in the Census of Housing. R. L. Ragatz (1970) has shown that in the late 1960's 3 million families in the United States (5 per cent of the total) owned "vacation homes" narrowly defined as buildings "originally constructed for the purpose of leisure-time activi-ties". The number would be considerably greater if converted farmhouses and fishing cottages were to be included. In the late 1960's vacation homes made up 5 per cent of the total housing stock in the United States. Ten per cent of new residences being constructed each year were second homes. Numbers were rising

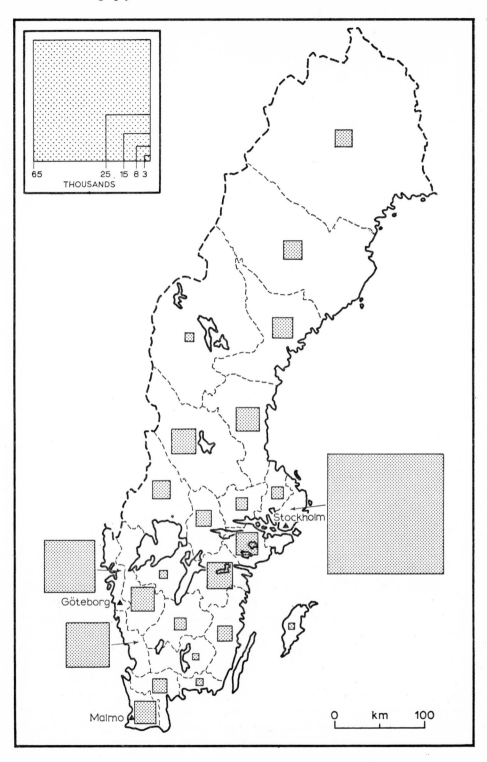

FIG. 5.4. Second homes in Sweden, 1962.

by between 100,000 and 200,000 annually, and could reach 300,000 by the late 1970's.

It is not easy to elucidate national differences in rates of second-homes acquisition. In addition to differences in car ownership, lengths of holidays with pay, general affluence, and the occupation of caravans (which provide alternative seasonal accommodation), one might draw attention to variations in urban living conditions. Many city dwellers in Continental Europe live in high-density apartment buildings and seek compensation at weekends in the form of a rural second home. This is perhaps less likely to be the case of many urban dwellers in Britain, who occupy semi-detached houses with individual gardens and thus possess a fragment of open space in which to spend their leisure time.

Patterns of Development

At a high level of generalization the proliferation of second homes through time might be fitted to a concentric-ring model around urban demand centres. Limited opportunities for personal mobility prior to the automobile age restricted the establishment of second homes to areas only short distances beyond the edge of the built-up city yet which were sufficiently distant to allow urban dwellers to escape to a "non-urban" environment. Rings of summer suburbs were identified close to many major European cities in the early 1900's. A detailed investigation of this phenomenon around Lyons has shown how the rings have broadened through time as transport technology moved from the era of the horse-drawn carriage before World War I to the age of the fast and comfortable private car in the 1950's and 1960's. The same study demonstrated the internal dynamism of the concentric-ring model, as continuously built-up suburbia expanded outwards to engulf areas which had previously contained many second homes, and as more distant settlements, which had functioned as seasonal suburbs between the wars, became metropolitan villages for commuters.

A comparison of various case studies shows marked spatial differences in the breadth of second-home hinterlands at any one period in time. Not surprisingly, larger cities generating greater demands for second homes, command broader hinterlands than smaller cities. In addition, there is a contrast between the proximity of second homes used for weekends around European cities and their far more widespread distribution in North America and Australia. Thus in the early 1960's two-thirds of all Swedish second homes were sited under 50 km from their owners' primary residences, and in Denmark four-fifths were less than 65 km away. G. P. Wibberley (1971) has noted a similar proliferation of second homes close to the built-up margins of British cities. The same is true around east European cities, e.g. in the Buda Hills only a few kilometres from the centre of Budapest. By contrast with this very restricted scale of weekend mobility, far greater distances appear to be acceptable in the broader and more highly motorized environments of North America and Australia, where up to 320 km between first and second homes is accepted as a reasonable drive for a weekend. Such national differences are linked to variations in rates of car ownership, of automobile comfort, and the existence of motorways and freeways permitting fast driving speeds.

One might also argue that there are national contrasts in the perception of distance, between small European countries, with a variety of closely interwoven environments juxtaposed over each relatively small section of terrain, and the much larger countries of the New World, where less diversified *milieux* are encountered.

The concentric-ring pattern provides a useful generalization to describe the evolution of second-home acquisition through time. Its major weakness is that it assumes an homogeneous peri-urban environment uniformly endowed with sites for second-home establishment. This is far from being the case. Hinterlands are made up of series of sectors and points with differing conditions of access, and are thus of varying suitability for second homes. In addition, urbanites perceive points as "attractive" for large variety of reasons linked to each area's physical geography and cultural attributes.

Real-world patterns of second-home distribution show a strong relationship to main roads and railways, fanning out from urban demand centres. In addition, some types of site condition are considered by urbanites to be more attractive than others. Hilly areas with varied landscapes, for example, are more favoured than broad, treeless plateaux. Water surfaces, both natural and man-made, are highly "attractive" features, and thus large numbers of second homes are found along rivers, coasts, natural lakeshores, and the margins of reservoirs. Other important factors affecting distributions of second homes include the existing settlement pattern (in other words, spatial variations in the supply of property for conversion) and local differences in planning policy regarding the construction of estates of new second homes.

An attempt to integrate factors affecting the distribution of second homes has been made in Sweden by H. Aldskogius (1969) who formulated the concept of "recreational place utility" which is a measure to summarize individuals' ranking of all sites from which they might choose to establish second homes. Both site (landforms, surface water, vegetation, land-use, and service facilities) and situation (roads from the demand centre into the rural hinterland, public transportation) characteristics have to be evaluated and ranked throughout the study area. Aldskogius assumed that if two sites of equal attractiveness were available, the potential second-home owner would choose the site closer to his primary residence.

This concept has considerable validity in depicting spatial aspects of second-home acquisition, but it also has limitations. First, country cottages are inherited by city dwellers in many parts of the world where rural/urban migration has been relatively recent. Heirs are not, therefore, confronted with locational choices but whether rural houses should be used as second homes or disposed of in some other way.

Second, a rural area might be selected as attractive for its social rather than its landscape features. Specific areas might appeal because of their sentimental attractiveness to urbanites. They might, for example, comprise the villages from which one's family migrated in the past. One might just "like the local people" or find the social or cultural atmosphere agreeable. French Protestants, for example, in a predominantly Roman Catholic country, find such an attraction in Protestant parts of the southern Massif Central.

Third, man-made landscape features, such as the recreational lakes in minor valleys in Mid-West of the United States, may be considered attractive. They owe relatively little to local conditions of physical geography but much to the enterprise and initiative of landowners or local authorities.

Fourth, increasing proportions of second homes are being built on special estates. The relative attractiveness of such estates depends, *inter alia*, on the effectiveness of advertising.

Finally, land-use planning controls are becoming more rigorous. They will condition patterns of second-home establishment to a far greater degree than in the past.

Second-home provision has become an important branch of the real-estate business in many developed parts of the world. The demand for second homes now stems from urbanites without rural "roots" rather than from the existence of rural/urban links that have been fostered through townward migration as was the case in the past. Variations in the fashionability of areas, resulting from the differential effectiveness of advertising and the diffusion of information between actual and aspiring second-home occupiers, will become increasingly important in the future.

Effects of Seasonal Suburbanization

One assumes that city dwellers benefit from their chosen stay in second homes. R. L. Ragatz (1970a, b), writing in the North American context, has suggested that the possibility of expanding the second-home market as an alternative to land uses which are either obsolete (farming and mining) or difficult to attract to the countryside (manufacturing and commerce) would be a rational alternative for the economic advance of backward rural regions, such as Appalachia and northern parts of New England. Positive advantages of second-home development include opportunities to sell off surplus land and buildings at the higher prices that urbanites are willing to pay; for rural authorities to obtain additional taxes; and for business to be increased in local shops, cafés, garages and builders' firms. C. Jacobs, planning officer for the Welsh county of Denbighshire, estimates that cottage buyers in Wales contribute about £4 million to the economy each year through the purchase of goods and services and provide jobs for about 15,000 people (*The Times*, 2 March 1972). He adds that local people do not want the isolated and run-down cottages most favoured by cottage hunters who have a strong urge for creative restoration.

Disadvantages also result. Rural driving conditions may become hazardous as country roads are congested by second-home traffic driven at fast speeds with urban driving manners. Agricultural land may be invaded by second homes and their gardens located in positions that hamper programmes for farm restructuring. The visual environment may deteriorate if do-it-yourself second homes are constructed which become little more than rural slums. The scattering of second homes across hillsides will enrage landscape preservationists. Even the building styles of modern second homes may be quite inappropriate to some rural areas. In Wales the emphasis is on converting existing property rather than building new cottages. The "cottage rush" of recent years has provoked a strong feeling among some Welshmen that the cumulative effect of this erodes the distinctive social and cultural fibre of rural Wales.

Reactions to second-home establishment tend to vary between peri-urban stretches of countryside and more remote areas. Detailed investigations in France have emphasized the disadvantages which arise when second-home owners flood into peri-urban areas such as the Paris Basin. Prices of land and houses shoot up and become quite beyond the scope of farmworkers seeking to acquire a cottage or small farmers trying to enlarge their holdings. Landlords are much keener to sell property, and thereby make a profit, than to rent out land or buildings. If sales are handled by urban estate agents a proportion of the profit will not even reach the rural inhabitants. A new type of settlement geography is developing around major cities as the number of weekenders increases and the permanent population declines. Such a contraction hastens the disappearance of village schools and other services which, in turn, encourages further outmigration of full-time rural residents.

Weekenders generally purchase their provisions before they leave the city. They buy little at village shops, which may be unprepared to meet urbanites' demands, and are certainly unable to satisfy them at supermarket prices. Temporary residents demand piped water supplies and mains drainage. Very often the bulk of the costs incurred in their provision have to be met by the permanent residents. Both local inhabitants and second-home residents seek the services of local builders and workmen. Neither group is satisfied. City dwellers find rural workers slow and inefficient. Rural residents complain that garagemen and builders prefer to cater for the affluent urbanite and ignore pleas for help with farm machinery. Some new forms of employment are created following the establishment of second homes, e.g. for gardeners and part-time housekeepers. Builders are kept busy renovating cottages and building new second homes, but their lorries damage country roads. Repair charges have to be met largely by permanent residents.

By contrast with the situation in peri-urban areas, advantages derived from second-home establishment seem to outweigh disadvantages in remote and economically impoverished rural areas. Second-home occupants often stay for long periods each summer. Three-quarters of village mayors interviewed in the French Massif Central found that advantages predominated. They listed increased trade for local shopkeepers as the most important benefit. The second most important feature was the seasonal increase in activity and social life which the presence of second-home occupants provoked. The general points regarding the social and economic impact of catering for visitors in the countryside, which have already been discussed, apply equally well in the context of second homes.

Planning Problems

Serious problems of rural management have arisen in areas where large unplanned concentrations of second homes have developed across limited stretches of terrain. V. Gardavsky (1969) has described the situation in seasonal suburbs close to Prague where the pollution of the atmosphere and surface water has become noticeable in recent years, partly because of the presence of second homes with inadequate mains drainage. Attractive environments which were selected for second-home construction in the past have become saturated. "The territories with optimal recreational parameters are gradually changed into areas that are quite unsuitable for recreation" (p. 19). Because of the high density of second homes in some areas, "it follows that they represent a type of environment which does not differ very much in its quality from a town setting" (p. 24). Other problems of saturation have been experienced along Scandinavian lake-shores and coasts, where second homes are conflicting with the demands for day recreation space, particularly for bathing beaches, and are adding to traffic congestion. Attempts at land-use zoning have been started in Denmark, Norway, and Sweden to delimit areas where second homes might be developed in the future and others where they should be prohibited, especially along valuable stretches of coast and in the mountains.

Rising demands for second homes will require effective management schemes for the provision of roads, electricity, piped water, and mains drainage. Strategies for concentrating second homes on estates offer the most satisfactory solution from the planners' point of view. Concentration would mean that schemes for farm enlargement and afforestation would not be impeded and rural landscapes would not be "eroded" by a scatter of second homes. Only the immediate environment of the second-home estates would have to be sacrificed. One must anticipate that specially constructed estates—almost micro-towns—which are already well established in North America, will become more widespread in the future in other parts of the developed world. As well as spreading urban ways of life into the countryside it would seem that seasonal suburbanization will take on a new form and dimension far removed from the existing pattern of rural settlement. In such an environment of change efficient land-use planning will be essential.

REFERENCES AND FURTHER READING

Leisure and the Countryside

The report on leisure submitted to the Strasbourg Conference of European Conservation Year is presented by:

HOOKWAY, R. J. S. and DAVIDSON, J. (1970) *Leisure: problems and prospects for the environment*, Countryside Commission, London.

Rising rates of car ownership are discussed in:

RODGERS, H. B. (1971) The sociological trends towards greater leisure, car ownership and car use, *Ekistics* **184,** 216–19.

Recreational travel is considered in:

COLENUTT, R. J. (1969) Modelling travel patterns of day visitors to the countryside, *Area*, **2,** 43–47.

The impact of the recreation industry on rural residents is discussed by:

SAVILLE, J. (1966) Urbanization and the countryside, in HIGGS, J. (ed.), *People in the Countryside: studies in rural social development*, National Council of Social Service, London, pp. 13–34.

A detailed sociological discussion of rural/urban contacts through catering for visitors in parts of France is included in:

RAMBAUD, P. (1967) Tourisme et urbanisation des campagnes, *Sociologia Ruralis* **7,** 311–34.
RAMBAUD, P. (1969) *Société rurale et urbanisation,* Seuil, Paris.

Commentaries on the impact of tourism in the countryside are provided by:

ANON. (1971) *Submission to Defence Lands Review Committee on the Conservation of Natural Beauty and the Use of the Countryside for Open-air Recreation,* Countryside Commission, London.
BRACEY, H. E. (1970) *People and the Countryside,* Routledge & Kegan Paul, London.
COSGROVE, I. and JACKSON, R. (1972) *The Geography of Recreation and Leisure,* Hutchinson University Library, London.
DOWER, M. (1972) Amenity and tourism in the countryside, in ASHTON, J. and HARWOOD LONG, W. (eds.), *The Remoter Rural Areas of Britain,* Oliver & Boyd, Edinburgh, 74–90.
LAW, S., BURTON, T. L., CHERRY, G., and KAVANAGH, N. (1970) *The Demand for Outdoor Recreation in the Countryside,* Countryside Commission, London.
MAW, R. (1971) Construction of a leisure model, *Ekistics* **184,** 230–8.
MORGAN-JONES, J. (1972) Problems and objectives in rural development board areas, in ASHTON, J. and HARWOOD LONG, W. (eds.), *The Remoter Rural Areas of Britain,* Oliver & Boyd, Edinburgh, 109–29.
PHILLIPS, A. A. C. (1970) *Research into Planning for Recreation,* Countryside Commission, London.

Second Homes

Interesting predictions of planning needs in the future are presented by:

WEBBER, M. M. (1968) Planning in an environment of change: Part I, Beyond the industrial age, *Town Planning Review* **39,** 179–95.

The classic study of second homes is presented by:

WOLFE, R. I. (1951) Summer cottagers in Ontario, *Economic Geography* **27,** 10–32.

A broad discussion of second homes as historic and contemporary phenomena is contained in:

CLOUT, H. D. (in press) The growth of second-home ownership: an example of seasonal suburbanization, in JOHNSON, J. H. (ed.), *The Geography of Suburban Growth,* Wiley, London.

The concept of recreational place utility is formulated in:

ALDSKOGIUS, H. (1969) Modelling the evolution of settlement patterns: two studies of vacation house settlement, *Geografiska Regionstudier* **6,** 1–108.

Second homes are discussed in various national contexts:

Australia
MARSDEN, B. S. (1969) Holiday homescapes of Queensland, *Australian Geographical Studies* **7,** 57–73.

France
CRIBIER, F. (1966) 300,000 résidences secondaires, *Urbanisme* **96–97,** 97–101.
CLOUT, H. D. (1969) Second homes in France, *Journal of the Town Planning Institute* **55,** 440–3.
CLOUT, H. D. (1970) Social aspects of second-home occupation in the Auvergne, *Planning Outlook* **9,** 33–49.
CLOUT, H. D. (1971) Second homes in Auvergne, *Geographical Review* **61,** 530–53.

Sweden
ALDSKOGIUS, H. (1967) Vacation house settlement in the Siljan region, *Geografiska Annaler* **49B,** 69–96.
NORRBOM, C. E. (1966) Outdoor recreation in Sweden, *Sociologia Ruralis* **6,** 56–73.

United Kingdom

BARR, J. (1967) A two-home democracy? *New Society* 313–15.

JACOBS, C. A. J. (1972) Second homes in Denbighshire, *County of Denbigh: Tourism and Recreation Report* **3,** 1–60.

MARTIN, I. (1972) The second-home dream, *New Society*, 18 May 1972, 349–52.

WIBBERLEY, G. P. (1971) Rural Planning in Britain: a study in contrast and conflict, inaugural lecture at University College London, delivered 27 May 1971.

United States

RAGATZ, R. L. (1970a) Vacation housing: a missing component in urban and regional theory, *Land Economics* **46,** 118–26.

RAGATZ, R. L. (1970b) Vacation homes in the northeastern United States: seasonality in population distribution, *Annals Association of American Geographers* **60,** 447–55.

TOMBAUGH, L. W. (1970) Factors influencing vacation home location, *Journal of Leisure Research* **2,** 54–63.

WOLFE, R. I. (1970) Vacation homes and the gravity model, *Ekistics*, **29,** 352–3.

An interesting example of second homes in Eastern Europe is discussed by:

GARDAVSKY, V. (1969) Recreational hinterland of a city taking Prague as an example, *Acta Universitatis Carolinae (Geographia)* **1,** 3–29.

CHAPTER 6

LAND-USE PLANNING

JOHN WELLER (1967) aptly remarked that "rural planning is as Cinderella to her sisters town planning and economic planning" (p. 93). One might argue that his comment is valid throughout the developed world, since members of the planning profession have been occupied, on the one hand, with urbanization and all its attendant problems, and, on the other, with achieving desired rates of economic growth at both the national and regional scales. In the inter-war years it may have been reasonable to think of the countryside as an environment to be set aside for food production and protected from urban encroachment. As the years have passed, increasing numbers of highly mobile, car-owning urbanites have exerted new pressures on the countryside for residence and recreation. In the 1970's the town and the countryside cannot be viewed satisfactorily as discrete *milieux* for environmental management.

This situation has been recognized with varying degrees of realism in the planning frameworks of individual countries and regions over the last quarter of a century. One manifestation of this trend is found in the increasing attention being paid to amenity issues and to the management of rural land in such ways as to cater for the rising demands coming from city dwellers for recreation in the countryside. The emphasis is now shifting from restrictive, protectionist policies to more positive proposals for development and conservation. However, in the eyes of some professional planners such a developmental approach has not yet made sufficient progress, and the last quarter century has formed wasted years.

Powers and techniques of land-use planning, together with policies for managing the countryside, vary greatly from country to country. The following discussion will concentrate on English experience, considering, first, the legislative background, and, second, the designation and management of recreation areas.

THE LEGISLATIVE BACKGROUND

Country planning was coupled with urban planning in England and Wales in a Town and Country Planning Act dating from 1932, but not until after World War II was an effective planning mechanism devised. In the 1930's the Government had urged Sir Montague Barlow to investigate the implications of the increasing concentration of England's population in the urbanized "hour-glass" zone orientated north-west to south-east from Lancashire to the London region.

A stream of ideas involving rural management came from the reports of the Barlow Commission (1940) and the Scott Committee (1942). The latter sought to create a centralized planning body; to encourage rural industry and commerce; to maintain prosperous farming; to resuscitate village and country life; and to preserve rural amenities. In the same year, the Uthwatt Committee went even further in arguing for the establishment of effective machinery to implement land-use planning.

The general attitude of the Scott Committee's report was that farming should retain a prescriptive right over land and that urban and other uses would be permitted to encroach only if it were clear that the land in question would not be required for producing food. Professor S. R. Dennison, in a minority report, urged that the agricultural industry should not be cushioned in such a privileged position. He maintained that changes in agricultural acreage, employment, and output needed to be contemplated. Four principles of rural management were put forward. First, all land in the countryside should be covered by the machinery of planning and no interests of national importance should be excluded. Second, much rural land would undoubtedly remain in agricultural use in the future, but this argument should not be used to justify the "preservation" of agricultural life and the exclusion of construction and development from the countryside. Third, the careful introduction of industrial activities into rural areas could be of great advantage to country dwellers. Fourth, the requirements of agriculture should be met through the normal channels for land use planning. Farming should not be given any prior rights over other, actual or potential, users of land.

Legislation which followed in the late 1940's established the basis for the present system of land-use planning in England and Wales. The Agriculture Act of 1947 aimed at a prosperous, stable, and efficient agricultural industry, and thus gave a slant in rural planning that was to the benefit of farmers. The Town and Country Planning Act (1947) for England and Wales made land-use planning compulsory over the whole country for the first time. It established that material alterations in the use of land could henceforth only be carried through after planning permission had been obtained. The English Act was consolidated by further legislation in 1962 and modified in 1968. Separate legislation exists for Scotland.

The planning authorities (counties and county boroughs in England and Wales) established in the 1940's were responsible for applying town and country planning legislation. It had, in fact, been the Act's intention to promote regional planning boards over and above the county level, but this was not achieved. Each planning authority was charged within a period of 5 years to prepare maps at various scales which set out broad proposals for land use in "the foreseeable future", namely about 20 years. Such development plans were to indicate the manner and stages by which the appropriate authority proposed that land would be used and land-use changes would result. Each plan would have to be reviewed at least once every 5 years. Development plans, together with any amendments, would have to be submitted to the (former) Minister of Housing and Local Government for approval after having been placed on deposit for public inspection. The Minister would then hold a public inquiry where objections to the plan could be raised. He would then approve the plan, or amendments, with or without modification.

The day-to-day implementation of development control is operated through the granting or refusal of permission for particular development schemes by the appropriate authority. Permission needs to be obtained for any change in land use technically recognized as "development". Certain modifications in land use, however, are not considered to be "development". These include minor alterations to buildings which do not change their external appearance; minor works linked to road improvement; and modifications in the use of land linked to agriculture and forestry. The final point is of particular relevance to this discussion.

The National Parks and Access to the Countryside Act (1947) gave local planning authorities in England and Wales additional powers to enhance the natural beauty of areas to be designated as national parks and to provide facilities for their enjoyment. In addition to the usual powers held under the Planning Act (for clearing unsightly development, discontinuing certain forms of land use, and preserving trees), the local authorities were also given powers to improve derelict land, plant trees, shrubs, and grass, provide access facilities to open land, appoint park wardens, and provide facilities for accommodation and meals along long-distance footpaths. Exchequer grants of up to 75 per cent of total costs are payable to local authorities for the provision of such facilities in areas designated by the National Parks (later the Countryside) Commission as national parks or as being of outstanding natural beauty. Full grants are available for the creation and maintenance of long-distance footpaths in England and Wales. The second part of the chapter considers in more detail the implementation of legislation for recreational planning in the countryside.

Special development controls operate in a variety of other situations. The Landscape Areas Special Development Order (1950) applies to defined areas of "natural beauty" in England and Wales. It provides that development for the purposes of agriculture and forestry (normally beyond the scope of development control) shall be subject to it, thus allowing local planning authorities additional powers of control. In addition, buildings of special architectural or historic interest which have been listed may not be demolished or altered without the consent of the local planning authority.

A Planning Advisory Group was set up in 1964 to review the existing planning system in England and Wales. The Group's report concentrated on problems associated with development plans and proposed that a distinction should be drawn between policy and strategic decisions, on the one hand, and detailed or tactical decisions, on the other. The Group urged that development plans submitted to the Minister for approval should deal only with the broad proposals and priorities affecting the area in question. Specific allocations of land for particular purposes and details of implementation would be the responsibility of local planning authorities. The Group criticized English planners for their apparent obsession for preparing detailed land-use plans. It urged them to review matters on a regional scale, so that schemes in contiguous planning authorities might be co-ordinated.

A new framework for planning was set up under the Town and Country Planning Act of 1968 following the submission of the Group's report. In the future, development plans would contain only major issues of policy. County plans would deal with the distribution of population, employment, major lines of communication, recreation, conservation, green belts, and overall development policies for towns and villages. These county plans would identify action areas where comprehensive planning was required. Local planning authorities would then prepare local plans to serve as guidelines for development control and provide a basis for more positive aspects of environmental management.

PLANNING FOR RECREATIONAL USE OF THE COUNTRYSIDE

Recreation, or the use of one's discretionary time, is a highly complicated concept involving a wide range of activities, taking place at varying times through the week and year, occurring with different degrees of intensity, and making a variety of demands on land resources. M. Clawson and J. Knetsch (1966) proposed a threefold classification of recreational space: from user-orientated areas, through intermediate areas, to resource-based areas. Figure 6.1 depicts the types of recreational space that might be distributed around a large concentration of urban population. User-orientated areas include city parks, playgrounds, and other areas within very easy access of central cities. Many of the areas in this category would fall beyond the scope of the present discussion. At the other extreme, resource-based areas comprise recreational environments that are far less "humanized". They include wilderness areas as well as national parks, which vary in conception from one part of the world to another. Normally speaking, resource-based areas are fairly remote from large concentrations of population. Considerable amounts of both time and money have to be expended by urbanites to reach such areas. Intermediate areas lie between the other two categories both spatially and in terms of their intensity of use. They need to be well located with respect to the user's place of residence, normally within an hour's driving time.

Early progress was made in England and Wales in providing municipal parks and other user-orientated recreation areas. Such provision has continued throughout the past 100 years. The designation of the other two types of recreation area has been a much more recent phenomenon, dating only from the post-war period. England and Wales were latecomers among the nations of the world in designating national parks. The first Canadian national park had been created at Banff as early as 1885, and since then more than two dozen other national parks have been set up in Canada. National parks were designated in the late nineteenth and early twentieth centuries in other parts of the New World. In the United States the vast Yellowstone Park, covering an area of virtually uninhabited forest land almost half the size of Wales, had been established in 1872. In 1916 a National Parks' Service was created to manage federal parks in the United States, "to conserve the scenery and natural and historic objects and the wildlife therein and provide for the enjoyment of the same in such manner and by such means as will leave them unimpaired for the enjoyment of future generations". National parks and nature reserves were designated in seven European countries in the first half of the present century before the first park was set up in England and Wales.

In 1929 the Addison Committee had been established to investigate the possi-

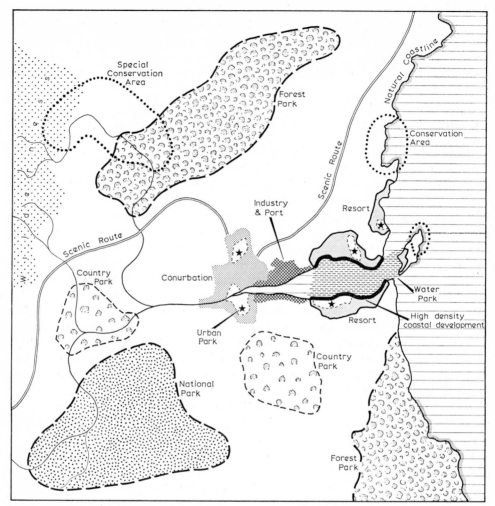

FIG. 6.1. Types of recreational space arranged around a conurbation.

bility of creating national parks in England and Wales and to consider the kinds of problem which might arise after their designation. The Committee presented its report in 1931. This favoured drawing up a national policy for parks, but no action was taken in the crisis decade of the 1930's. The matter rested until the Scott Committee Report (1942) which supported proposals for creating a central authority to delimit national park areas and recommended that a separate execu-tive body should both plan and control their use and development. Two years later a white paper on the Control of Land Use included proposals for setting aside land for national parks as part of national policy.

The recommendations included in the Dower Report on National Parks in England and Wales (1945) formed the basis of the Hobhouse Report of the National Parks' Committee (1947). This set out detailed proposals for a scheme of national parks and their administration and control. The Dower Report's definition of a "national park" was accepted. Each "national park should comprise an extensive area of beautiful and relatively wild country in which, for the nation's benefit, and by appropriate national decision and action the following should obtain: (i) characteristic landscape beauty strictly preserved; (ii) access and facilities for public open-air enjoyment amply provided; (iii) wild life and buildings and places of architectural and historic interest suitably protected; and (iv) established farming use effectively maintained."

The national parks were to be administered for the benefit of the nation, with the implication that "planning in park areas should not be carried out by the ordinary local government bodies with the National Parks Commission acting as an adviser and supplier of grants" (Cullingworth, 1967, p. 168). Similarly, the Hobhouse Report had urged for truly national administration: "If national parks are provided for the nation they should be clearly provided by the nation. ... Their distinct cost should be met from national funds." The legislation which resulted departed substantially from these recommendations.

The national parks were not envisaged as rural museums. In any case they would occupy areas that were used for farming and many other purposes including the accommodation of substantial numbers of residents. Progressive policies for management were to be implemented, "designed to develop the latent resources of the national parks for healthy enjoyment and open-air recreation to the advantage of the whole nation". Such management operations might include: the removal or "improvement" of inappropriate disfigurements (such as mineral workings); action to remove litter and prevent damage to crops, walls, and trees; tree-planting and replacement; assisting highway authorities to install parking places; providing holiday accommodation; and creating facilities for field studies. The legislation in 1949 which led to the creation of national parks charged the National Parks Commission with the almost contradictory tasks of both preserving and enhancing the natural beauty of park landscapes and also of providing opportunities for outdoor recreation.

The Hobhouse Report had listed twelve areas in England and Wales where national parks might be established. In fact only ten areas were designated (Table 14; Fig. 6.2). Recommendations for creating parks in the Norfolk Broads and on the South Downs were not carried through. Each of the areas selected fulfilled the following requirements: it was a zone of great natural beauty with a high potential value for open-air recreation and was of substantial continuous extent. Other requirements were added later. Park areas should have internal variety, and there should be at least one national park within fairly ready access of each of the main centres of population in England and Wales. This final aim has not been achieved. The nearest national park to London is in the Brecon Beacons. Certain rules were used in delimiting national park areas. Easily recognizable boundaries, such as roads and railways, were frequently followed since local-authority boundaries were rarely convenient. Settlements were not bisected by national parks' boundaries and an effort was made to exclude patches of unsightly development. By contrast, efforts were made to include scientific, historic, and architectural features on the margins of the proposed park areas.

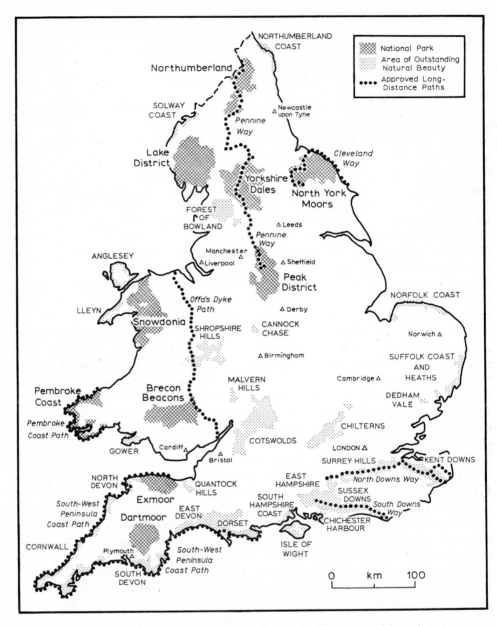

Fig. 6.2. National parks, areas of outstanding natural beauty, and long-distance footpaths in England and Wales.

TABLE 14. CREATION OF NATIONAL PARKS IN ENGLAND
AND WALES

1951	Peak District; Lake District; Snowdonia; Dartmoor
1952	Pembrokeshire Coast; North York Moors
1954	Yorkshire Dales; Exmoor
1956	Northumberland
1957	Brecon Beacons

Source: BRACEY, H. E. (1970), *People and the Countryside*,
Routledge & Kegan Paul, London, p. 235.

In spite of earlier proposals to the contrary, the Government decided that the
newly constituted planning authorities of the late 1940's should be given responsibility for national parks. A National Parks Commission (later the Countryside
Commission) was established, but its powers were largely advisory. As J. A.
Patmore (1970) has remarked: "administratively, the parks were scarcely
national in concept" (p. 196). Planning control remained in the hands of local
authorities, although the care of each park was vested in a local park committee.
Joint planning boards with executive powers independent of the county councils
were set up in the first two parks to be designated—the Peak and Lake Districts
(1951). Only in the Peak Park was a separate planning team established. Joint
advisory committees were created in the other parks which covered more than one
county (Brecon Beacons, Exmoor, Snowdonia, and the Yorkshire Dales). For
the four parks located in single counties (Dartmoor, Northumberland, North
York Moors, and the Pembrokeshire Coast), administration is through a committee of the appropriate county council. Two-thirds of the committee's members
are drawn from locally elected councillors and one-third appointed on the advice
of the National Parks Commission. Only the Peak and the Lakes authorities have
financial independence through power to precept on the county rates, though in
the case of the Lakes this is severely limited. Only the Peak Park employs its own
staff. All the others are staffed on a part-time basis by county officials. With the
exception of the Peak, and partially of the Lakes, national parks are run by local
authorities. In short, the administration of the ten parks, covering 9 per cent of
England and Wales, is a muddle and a negation of the national park ideal.

The National Parks and Access to the Countryside Act (1949) proved to be a
severe disappointment to those who had campaigned for national parks in the
period of government inaction between the wars. Successive reports had called
for a national parks system that was truly national, perhaps on the American
model, with a strong national agency having executive powers (including the
power to buy extensive tracts of land) and money to back up these powers. G. B.
Ryle (1969) has emphasized that in the so-called "national" parks "there is no
public ownership, no automatic public access and thus no positive management
plans or policies can be evolved for them" (p. 96). National parks can only be
conserved negatively by the operation of restrictive development controls. Habi-
tual rural pursuits and changes in agricultural and silvicultural land use are

unrestricted except by informal agreement between individual occupiers and the park authorities. "Just because national parks cannot be brought under any positive planned management for public use, they cannot be included in any detailed scheme where quantitative recreational need in the years ahead has to be assessed" (Ryle, 1969, p. 96).

As J. A. Patmore (1970) has noted: "Hamstrung by lack of finance and working through the normal process of development control, the administration of the parks could be but little different from that of surrounding areas. Most of the achievement has been negative rather than positive, but none the less valuable for that in terms of conserving the scenic heritage. The visitor will not see the damage which has been prevented" (pp. 198–9). But damage has taken place, none the less, and the environment of national parks has been violated as a result of the continuation of forms of land use that were established before the parks were designated. On Dartmoor, for example, permission for military training dates back to 1875 and has been renewed in 1948 and 1956. Firing ranges and military manoeuvres are hardly compatible with the basic aims of national parks.

In addition, permission for land-use changes has been given since the designation of national parks. Thus a ballistic missile early warning station has been built on the North York Moors, a nuclear power station in Snowdonia, and an oil refinery and an iron-ore stocking ground in Pembrokeshire. Potash is being mined in the North York Moors and china-clay working disfigures part of the Dartmoor landscape. In several of these instances the advice of the National Parks (or latterly Countryside) Commission has been overridden by ministerial decisions. Arthur Blenkinsop (1968) noted that: "Ministers gave assurances that the designation of an area as a national park meant that effective priority would be given to the protection of its beauty. . . . Each national park is its own monument to the weaknesses of such assurances" (p. 526).

Changes of an equally dramatic visual effect result from afforestation and modifications in agricultural land use. Such changes are quite legal and are beyond the direct control of the planning authorities as empowered under the 1947 Planning Act. Thus, for example, reclamation of moorland has proceeded apace in recent years in the North York Moors, Dartmoor, and Exmoor. The Exmoor Society documented the decline of the open moorland area from 23,500 ha in 1957–8 to 20,268 ha in 1965, representing an annual loss of 400 ha. Of this total, 360 ha were ploughed up for agricultural use and the remainder devoted to forestry. The plough-up campaign was encouraged, first, by the perfection of techniques for ploughing and re-seeding upland areas, and, second, by the fact that farmers were induced to plough marginal land by reclamation grants, available at the rate of £30 per ha under the 1967 Agriculture Act. Not only are the visual features of the national park being modified by ploughing, fencing, and tree-planting, but public access to areas of recreational space is reduced. The Countryside Commission in its report for 1969 summarized the general situation throughout the parks. "After 20 years the Parks are plainly vulnerable. To save them from spoliation will call for redoubled efforts on the part of ourselves, the Park authorities, and, not least, the public" (p. 1).

Individual parks are drawing up management schemes for their own areas.

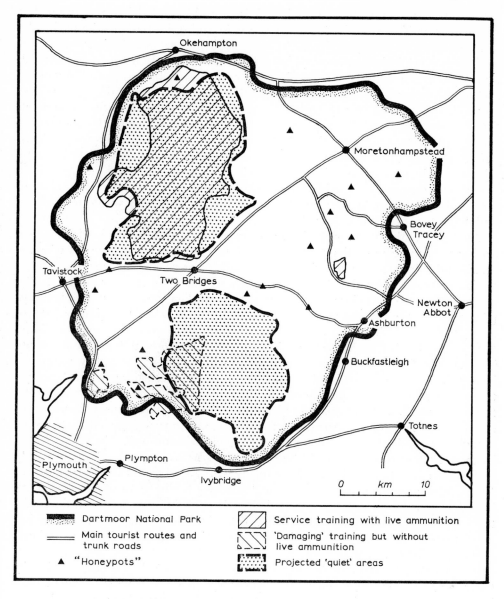

Okehampton

Moretonhampstead

Bovey
Tracey

Tavistock

Two Bridges

Newton
Abbot

Ashburton

Buckfastleigh

Totnes

Plymouth   Plympton

Ivybridge

0   km   10

Dartmoor National Park   Service training with live ammunition

Main tourist routes and   'Damaging' training but without
trunk roads   live ammunition

▲ "Honeypots"   Projected 'quiet' areas

Fig. 6.3. Dartmoor National Park policy plan, 1971.

Thus, for example, Dartmoor National Park Committee published a policy plan
in 1971. This recommended the zoning of two large areas for quiet pursuits
(Fig. 6.3) in which roads would be closed to motorists. A traffic management
scheme would be drawn up to direct visitors to the emptier picnic spots when the
well-known ones were full. The policy plan covers subjects such as the siting of
reservoirs, the working of minerals, afforestation, and the use of parts of the
national park by the Ministry of Defence. Comprehensive consideration of future

water extraction from the national park is desirable and the report urges that reservoirs should not be sited in the more wild and quiet parts of the park. The plan makes the necessary obeisance to the possible importance of mineral workings to the local and national economy, but stresses that restoration would be essential, and conservation should come first in quiet and conspicuous areas and in those of special character and scientific interest. It is strongly against the extension of china-clay extraction in the higher areas of open moorland in the south-west. Since afforestation and the military use of land are under review, the plan makes no direct recommendations, though it emphasizes the undesirability of the Ministry of Defence using land in a national park. Other measures suggested to cope with the estimated influx of 95,000 people on a peak day in the early 1980's, about twice that in 1967, include the creation of new footpaths, the designation of scenic and tourist routes, on some of which caravans and buses would be banned, and the control of mobile sales vehicles.

The Goyt Valley traffic experiment operated in part of the Peak District Park during the summer months of 1970 and 1971. The object was to assess whether "park and ride" could cope with traffic problems at rural beauty spots. A 6 km section of roadway, which had experienced serious congestion of visitor traffic in previous years, was closed to private cars at weekends and on public holidays (Fig. 6.4). Parking places were provided and visitors taken by minibus (free of charge) into the valley. They were encouraged to walk once they arrived. Some 45,000 people visited the valley whilst restrictions were in force, and on-site interviews and questionnaire surveys showed that closing the valley roads has been popular to an unexpected degree. The idea of a motorless zone in the countryside appealed to many people who left their cars and enjoyed walking along the traffic-free roads and way-marked paths, or riding in the minibuses. Reactions to charging for parking were generally favourable and surveys showed that visitors thought it would be reasonable to pay for use of the minibuses. The area affected by the park-and-ride system will be modified in the future and will operate only on Sundays and Bank Holidays. Monitoring will take place during the summer of 1972, and a permanent traffic order will not be established until the Peak Park Planning Board is completely satisfied that the right answer has been found for traffic management in the valley. This type of traffic management scheme has much to commend it, and might be considered for application in other national parks in the future.

In 1968 the Countryside Act replaced the National Parks Commission by the Countryside Commission. This new Commission has much wider concern than its predecessor. Its activities are no longer restricted to national parks. It is charged to keep under review all matters relating, first, to the provision and improvement of facilities for the enjoyment of the countryside, second, to the conservation and enhancement of natural beauty and amenity in the countryside, and third, to the need to secure public access to the countryside for the purposes of open-air recreation. In short, the emphasis of the Commission's powers has shifted from the preservation of the countryside to its use and conservation. Nevertheless, the administration of the national parks remains substantially unchanged.

During the 1960's most observers agreed that the separate planning authorities

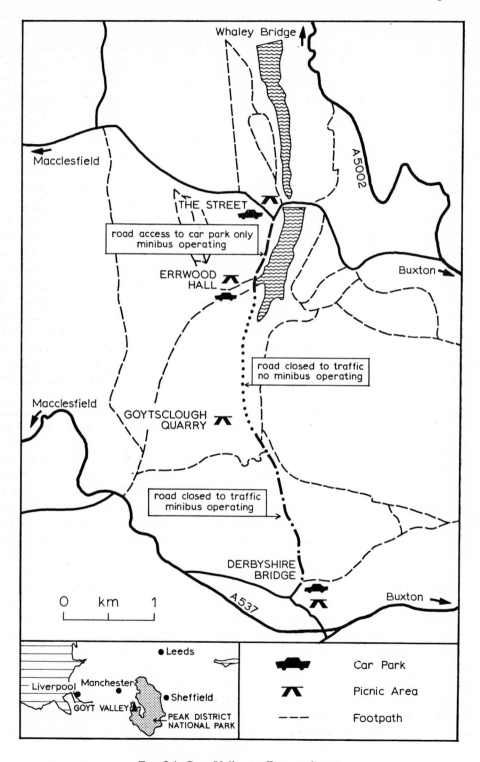

Whaley Bridge

Macclesfield

A 5002

THE STREET

road access to car park only
minibus operating

ERRWOOD
HALL

Buxton

road closed to traffic
no minibus operating

Macclesfield

GOYTSCLOUGH
QUARRY

road closed to traffic
minibus operating

DERBYSHIRE
BRIDGE

Buxton

O   km   1

A537

● Leeds

Liverpool Manchester

GOYT VALLEY

● Sheffield

PEAK DISTRICT
NATIONAL PARK

	Car Park
	Picnic Area
- - -	Footpath

FIG. 6.4. Goyt Valley traffic experiment.

charged with the management of the Peak and Lake Districts national parks had met with a far greater degree of success than had the county councils responsible for the eight other parks. Generally speaking, the county authorities had been reluctant to do much in the way of positive planning or to spend ratepayers' money. This situation was recognized by the Royal Commission on Local Government (Redcliffe/Maud report) in 1969 which recommended that each park should be administered by a financially independent authority employing its own staff. The white paper, entitled *Reform of Local Government in England* and produced by the Labour Government in 1970, took the same view. However, a second white paper presented by the subsequent Conservative Government one year later saw no need for a change in status. The Countryside Commission started a campaign to alter the Government's view, and in June 1971 it published a report by Sir Jack Longland on the future of national parks. This noted that the history of the national parks had been one of failing to live up to the ideals of the 1949 Act. Only the Peak Park and to a lesser extent the Lakes had been able, through their relative independence, to tackle the problems facing the parks. The Commission used the Longland report to support its argument that in future each national park should have its own autonomous planning board, including both local authority and nominated outside members, with power to appoint its own staff. The boards should include all the planning powers of the local planning authority and be able to precept the local authorities concerned for the costs they incur. Exchequer grants should be available to cover all costs save those incurred by the planning boards doing duties as local planning authorities. Such statements represent a repetition of the Hobhouse proposals of a quarter century ago.

This view was criticized on the ground that further *ad hoc* authorities would simply distort the total pattern of expenditure on recreation. Representatives of the County Councils' Association argued that national parks should remain firmly in the hands of local government. They objected to any diminution of the powers of their constituent members and insisted that the fact that there were farms and small industries, market towns, and villages within the national parks meant that the normal functions and service of local authorities were as relevant there as anywhere else. To remove the parks from local government would be prejudicial to local interests. The Welsh counties and other opponents to the Longland report maintained that national park planning boards would remove the planning powers and functions of local authorities over large areas of their territory and vest them in an undemocratically elected board, not directly answerable to the general electorate. It is interesting to note that two counties with direct experience of independent park authorities, Derbyshire and Westmorland, dissented from the CCA's line. The Government yielded to the CCA's policy, even though this meant rejecting the advice of its appointed official advisers on national parks, the Countryside Commission.

A very different point of view was taken by other commentators who argued that a strong case could be made for something along the lines of the American solution to create a strong central body, with money-spending powers, to buy and run pieces of the area itself. This would release the local authority from any worry about the financial consequences of the park apart from some beneficial side effects to local tourism.

In the summer of 1971 the Department of the Environment informed the Countryside Commission that its proposals contained in the Longland report were unacceptable. Faced with this rejection of their advice, the Commission sought a compromise with the CCA. This agreement, published in October, marginally improved the situation by proposing to do away with the multiplicity of committees in multi-county parks and to give each park a national park officer—though of undetermined status. This agreement fell short of the ideal on two key points. The parks would still be financed out of county council budgets, and their staff (who would not necessarily be full time) would still be county council employees.

The financial issue revolved around the fact that county councillors are elected to look after local interests. This means that they strive to keep the rates down. Unfortunately, about 40 per cent of all national park expenditure was rate-borne. In the financial year 1970/1 the total expenditure on all ten parks was under £900,000, less than one-quarter of what the Greater London Council spends on its parks and open spaces each year. Yet the national parks form the open spaces of the entire nation. The two parks with a measure of financial independence (the Peak and the Lakes) spent £395,000, 44 per cent of the total spent by all ten. The Peak, for example, spent £354 for every square mile of its territory. No other park reached £300, and the average expenditure per square mile in the other nine parks was only £150.

The second issue, namely staffing, involves the fact that running a national park ought to be a specialized business. Sensitive and dedicated planners are needed to protect landscapes, to provide information points and warden services to ensure that visitors derive the most from their trip without upsetting local interests or destroying wild life. Only the Peak has a staff of its own. Very curious situations have arisen in the other parks. In 1970 the Lake District Planning Board was apprehensive about plans for a bypass south of Keswick prepared by the Cumberland County Council. But the Board's officers responsible for the Keswick scheme were employed by the county council and could not, therefore, give objective advice. The Board had to employ an outside consultant who advised that the scheme would be wholly wrong from the point of view of the national park. Other examples could be cited to show the dilemma that county councils are placed in having to plan both for the conservation of national parks and for the general economic and social well-being of the rest of their counties.

The national parks face greater pressures of visitors in the future than ever before. The Exmoor Park Committee, for example, forecasts that the number of people within a 3½-hour journey of the park will nearly quadruple from just under 5 millions to nearly 19 millions as trunk-road improvements are carried through in the next ten years. To cope with numbers of this magnitude positive action is needed. Car parks and picnic places must be carefully sited and landscaped. If necessary, lanes must be closed to motor traffic and visitors must be asked to walk or use minibuses. Information services must be expanded. Land must be bought or access agreements made so that people can enjoy previously inaccessible landscapes. These things cost money—money on a scale that the county councils have failed to provide in the past and will be unable to provide in the future. The preservation of the integrity of the parks demands planning by full-time staff devoted to the management of the national parks.

In February 1972 Lord Sandford announced to the House of Lords that the Government would amend the Local Government Bill so that each national park would have a single body for administration, management, and planning. In every park there would also be a single chief officer with his own staff. Each park authority would be responsible for producing a single comprehensive plan and programme for action. Lord Sandford also promised much larger national contributions which would cover "the lion's share" of park expenditure.

This announcement represents a considerable advance for the Countryside Commission; however, those involved in the Campaign for National Parks (24 February 1972) were unsure of the precise implications of the wordage used by Lord Sandford. They stressed their belief in the need for truly independent authorities for each park. Future success will depend on the political engineering of the new park committees and the persuasiveness of their chief officers. Experience in the Peak District has shown that both can function successfully. Attitudes may, however, be different in areas further away from the direct influence of large cities, and strong opposition may arise to positive policies for developing the parks for the "healthy enjoyment and open-air recreation to the advantage of the whole nation" that was, after all, one of the key objectives of the 1949 Act.

In addition to resource-based areas in the form of national parks, intermediate recreation areas have also been designated as areas of outstanding natural beauty and, more recently, as country parks. The Dower and Hobhouse reports recognized that not all areas requiring special conservation treatment could be included in national parks' schemes. Hence they proposed that areas of landscape beauty should be defined and protected. The Hobhouse report had listed fifty-two conservation areas, but in fact the 1949 Act contained no special provision for the care of such areas. Instead it gave powers to the National Parks Commission to designate areas of outstanding natural beauty where Exchequer grants would be available on the same lines as in national parks. Areas of outstanding natural beauty thus received no special planning machinery and became the responsibility of local planning authorities charged with powers for "preservation and enhancement of natural beauty". By the autumn of 1971 more than twenty areas of outstanding natural beauty had been designated (see Fig. 6.2), representing over 6 per cent of the area of England and Wales. In fact, local planning authorities have been reluctant to use their powers for improvement schemes in spite of Exchequer grant aid. This has been due to an unwillingness to incur expense and a general failure to think in terms of catering for visitors. In autumn 1971 John Cripps, chairman of the Countryside Commission, noted that in the future the Commission would concentrate on improving the administration of existing areas of outstanding natural beauty rather than designating new ones.

In 1966 the White Paper on *Leisure in the Countryside in England and Wales* outlined a formula for the creation of country parks. Such parks would make it easier for town dwellers to enjoy their leisure in the open without having to travel great distances and thereby add to congestion on the roads. They would alleviate pressure on more remote and solitary places. In addition, they would reduce the risk of damage which results when people simply seek a place to spend a while in the countryside.

A country park was described in broad terms in the Countryside Act (1968) as "a park or pleasure ground for the purpose of providing, or improving, opportunities for the enjoyment of the countryside by the public". The concept was deliberately vague but, in reality, the Countryside Commission has suggested that country parks should be readily accessible to motor vehicles and pedestrians and should be not less than 10 ha in size, although "the area as such is less important than the capacity to absorb a considerable number of people or to provide a variety of recreational activities". The Commission has outlined three basic priorities for grant aid for country parks. One-quarter of the costs incurred must be covered by the local authority or private agency concerned, but up to 75 per cent can be covered by Exchequer grants. Priority in grant allocation should be linked to the three following conditions: first, for areas where existing recreation facilities were inadequate (close to large concentrations of population); second, for the improvement of areas already in use for recreation which could be converted to country parks with only modest expenditure; and, third, in areas of derelict or under-used land, particularly where it was owned publicly.

The first few country parks in England and Wales demonstrate the diversity of this category of recreation facility. The Wirral Way in Cheshire has involved the conversion in a popular recreation area of 13 km of disused railway line for walking and riding, with views across the Dee estuary. Car parks and picnic areas will be provided. The Elvaston Castle country park in Derbyshire has conserved an area of agricultural land, woods, parkland, and formal gardens. Beacon Fell in Lancashire is an isolated hill in the Forest of Bowland which was partly planted up with timber as a water-gathering ground. Lancashire County Council are providing access roads and car parks on the site. The Cotswold Water Park will provide facilities for water recreation, picnicking, walking, and fishing in an area of worked-out gravel pits in the Upper Thames Valley. By the autumn of 1971 £1 million in grants had been given toward the cost of setting up more than forty country parks throughout the country, with grants promised for a further thirty-six (Fig. 6.5). In addition, £200,000 in grants had been approved for establishing thirty-nine picnic sites, with another fifty awaiting consideration.

PROTECTION OR DEVELOPMENT?

In the immediate post-war period, land-use planning policies in England and Wales sought to tackle the pressing urban problems of the day. Long-term issues took second place. The countryside, all too often, was seen as an environment which should be protected. Thus the management of rural land under the 1947 Act aimed to shelter the countryside from detrimental "development", but, nevertheless, statutory authorities, such as the Services, forestry, roads, railways, gas, water, and electricity, largely escaped the realm of local planners. The planning machine paid inadequate recognition to the newly acquired mobility of

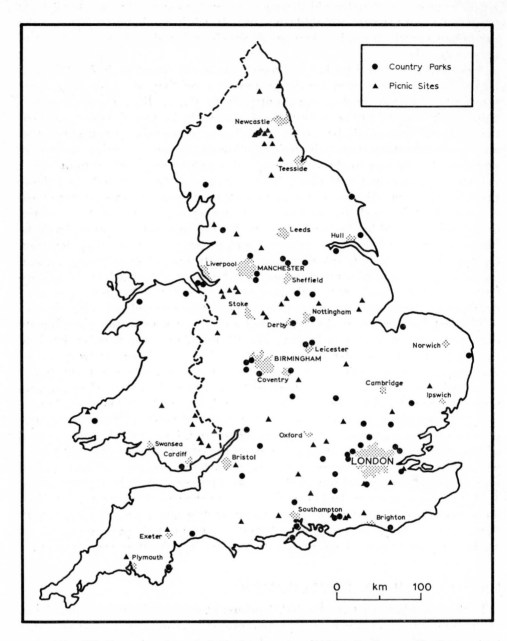

Fig. 6.5. Country parks and picnic sites recommended by the Countryside Commission and approved by the Department of the Environment/Welsh Office for grant aid, January 1972.

urbanites and their growing demands on the countryside. As H. E. Bracey (1970) explained: "protection has in the past been the conventional way of safeguarding rural amenities, but the interests of the farmer and the town dweller seeking to enjoy rural amenities are often sharply opposed, for agricultural processes normally require enclosure whereas active recreation in the countryside demands access" (p. 126).

In spite of its many limitations, G. P. Wibberley (1970) has shown that protective rural planning will remain important in at least three respects in the future. Large sections of rural Britain, and especially the better grades of land, will be needed to fulfil the food requirements of this country. The precise area will depend on a great variety of factors, including international trade commitments and the rising productivity of domestic agriculture. Second, changing pressure from within the rural economy and city-derived pressures from outside will inevitably lead to changes in the countryside. There is a need to "integrate and control these pressures so that they do not badly disrupt the productivity and amenities of these rural environments" (p. 287). Third, the countryside will gain much greater importance in the future as a vital compensation to the normal living environment of the city. It must be allowed to remain sufficiently distinctive in terms of landscape and economy to guarantee such compensation. Writing of urban dwellers and the continuing need for preservation, Wibberley (1971) notes: "It may not matter to these people that the purpose and occupants of this rural settlement pattern have changed greatly and bear little relationship to the real needs for people and settlement of modern-day agriculture, forestry, rural recreation, and rural industries. It is enough for them that the rural settlement pattern has both the appearance and the buildings of an earlier rural civilization" (p. 361).

Nevertheless, Wibberley (1970) has shown that rural planning demands much more than simply implementing policies for protection. Numerous problems may be identified where a developmental approach would be more appropriate. R. J. Green (1971) has offered a similar appraisal of planning operations in rural Britain: "In the negative sense that they have prevented development which might have intruded into the countryside, and have encouraged building where it seemed to do least harm, rural planning policies have been moderately successful, but the 24 years since the Town and Country Planning Act of 1947 represents 2 decades of wasted opportunity for positive rural planning" (p. 3). In Green's view, local planning authorities have made little attempt to draw up policies which, first, recognize the changing economic and social conditions of rural areas, and then take advantage of the increasing physical and social mobility of the rural population.

The effective implementation of developmental planning in rural areas requires following a series of essential steps. Management proposals need to be positive, progressive, and comprehensive. They must recognize that numerous and sometimes conflicting interests are involved in the use, conservation, and management of the resources of the countryside. Special finances are needed for rural planning. They should be devoted to schemes that will cater for likely future demands on rural resources (for food production, forestry, recreation, residence, and so on)

rather than to programmes which simply subsidize and maintain the *status quo*. In spatial terms it is necessary to recognize that the existing rural settlement pattern should be modified to satisfy changes in both mobility and occupation and also to accommodate new ways of using rural resources. The chapters which follow will return to many of these themes.

REFERENCES AND FURTHER READING

The background to land-use planning in a British context is provided in:

BRACEY, H. E. (1970) *People and the Countryside*, Routledge & Kegan Paul, London.
CULLINGWORTH, J. B. (1964) Planning for leisure, *Urban Studies* **1**, 1–25.
CULLINGWORTH, J. B. (1967) *Town and Country Planning in England and Wales*, Allen & Unwin, London.
WELLER, J. (1967) *Modern Agriculture and Rural Planning*, Architectural Press, London.

Problems arising from the recreational use of the British countryside are considered in:

BURTON, T. L. and WIBBERLEY, G. P. (1965) Outdoor recreation in the British countryside, *Wye College Studies in Rural Land Use* 5.
CLAWSON, M. and KNETSCH, J. C. (1966) *Economics of Outdoor Recreation*, Resources for the Future Inc., Johns Hopkins Press, Baltimore.
COPPOCK, J. T. (1966) The recreational use of land and water in rural Britain, *Tijdschrift voor Economische en Sociale Geografie* **57**, 81–95.
COUNTRYSIDE COMMISSION (1969) *Annual Report*, HMSO, London.
CRACKNELL, B. (1967) Accessibility to the countryside as a factor in planning for leisure, *Regional Studies* **1**.
DOWER, M. (1970) Leisure: its impact on man and the land, *Geography* **55**, 253–60.
MERCER, D. C. (1970) The geography of leisure, *Geography* **55**, 261–73.
PATMORE, J. A. (1970) *Land and Leisure*, David & Charles, Newton Abbot.
RODGERS, H. B. (1969) Leisure and recreation, *Urban Studies* **3**, 368–84; also in *Developing Patterns of Urbanisation*, Oliver & Boyd, Edinburgh (1970).
SELF, P. *et al.* (1965) A policy for countryside recreation in England and Wales, *Town and Country Planning* **33**, 473.

Valuable up-to-date information on recreation planning in Great Britain is contained in copies of the *Recreation News Supplement*, published by the Countryside Commission. International studies of the recreational use of rural land include:

BURTON, T. L. (1966) Outdoor recreation in America, Sweden and Britain, *Town and Country Planning* **34**, 456–61.
SIMMONS, I. G. (1966) Wilderness in the mid-20th century USA, *Town Planning Review* **36**, 249–56.
SIMMONS, I. G. (1967) Outdoor recreation as a land use in the USA, *Tijdschrift voor Economische en Sociale Geografie* **58**, 183–92.

Information on national parks in England and Wales is found in:

BLENKINSOP, A. (1968) The National Parks Commission, *Town and Country Planning* **36**, 525–6.
BOARD, C. *et al.* (1970) People on Dartmoor, *Geographical Magazine* **42**, 266–79.
BRUNSDEN, D. (1971) Plan for the character of Dartmoor national park, *Geographical Magazine*, **44**, 209–10.
CAPNER, G. (1969) Exmoor: a response to change, *Town and Country Planning* **37**, 262–9.
DARBY, H. C. (1963–4) British National Parks, *Advancement of Science* **20**, 307–18.
EDDISON, T. (1971) National parks reform, *Town and Country Planning* **39**, 416–18.
HALL, D. (1972) National Parks, *Town and Country Planning*, **40**, 124–6.
HALL, P. (1972) Parks, *New Society* **24** (2), 396.

JACKSON, R. (1970) Motorways and national parks in Britain, *Area*, **4**, 26–29.

LONGLAND, J. (1971) *Reform of Local Government in England and Wales*, Countryside Commission, London.

RYLE, G. B. (1969) Three kinds of parks, *Town and Country Planning* **37**, 94–98.

WHITE, J. (1970) The provision of country parks: a critique, *Town and Country Planning* **38**, 289–92.

Policies for the protection and the development of the countryside are discussed in:

GREEN, R. J. (1971) *Country Planning*, Manchester University Press, Manchester.

WIBBERLEY, G. P. (1970) Rural planning in Britain: protection or development, *Journal of the Town Planning Institute* **56**, 285–8.

WIBBERLEY, G. P. (1970) Rural conservation, *Chartered Surveyor* **103**, 76–81.

WIBBERLEY, G. P. (1971) Our green jelly, *Town and Country Planning* **39**, 360–2.

CHAPTER 7

STRUCTURAL CHANGES IN AGRICULTURE

HIGH proportions of the land surfaces of the countries in the developed world will still be in agricultural use in the final quarter of the twentieth century and beyond in spite of the abstraction of land for urban growth, afforestation, and other non-agricultural purposes. To be efficient, modern farming requires a thoroughgoing remodelling of fields, farm sizes, farmsteads, land-use traditions, and other rural structures that have been inherited from the past. The enclosure movement in parts of Great Britain swept away the medieval structure of open fields and highly fragmented plots of land, replacing them by large, rectangular fields with additional isolated farmsteads away from the old settlements, the whole pattern being bound together by new systems of roads. Other enclosure movements produced fundamental changes in some other parts of northern Europe, but these were the exceptions rather than the rule. Enormous structural problems remain in much of non-Communist Europe. This chapter will review a variety of types of restructuring operation and the subsequent changes that have been produced in rural landscapes over recent years.

STRUCTURAL CHANGES IN GREAT BRITAIN

Modern farm management requires larger holdings, larger fields, new roads, and buildings to suit an approach to capital investment, production, and mechanization which differs in degree from that at any period in the past. J. Weller (1967) has seen this as "perhaps the greatest change in the social fabric of Britain since the breakdown of the feudal system" (p. 215). By the end of the twentieth century, fields in eastern England will be measured in hundreds of hectares whereas fields of 16–20 ha are now considered to be large, indeed almost twice the size of the average *farm* in the Common Market countries. Farm enlargement involves not only a reorganization of the field layout but also changes in the farm's road system, improvements in fertilization and land drainage, and perhaps the centralization of buildings to suit new management techniques. Old farm buildings may become redundant and uneconomic to retain.

Hedgerows and other micro-features of the rural landscape are being swept away in this structural transformation. The English enclosure movement in the eighteenth and nineteenth centuries involved planting perhaps 290,000 km of hedge to surround fields of 2–4 ha in size. G. R. Allen (1972) has shown that the hedgerows and small fields were the consequences of special economic circumstances: cheap and abundant labour, whether hired or provided by the farming family; large seasonal variations in the work load with sharp peaks in the spring and autumn; and an agricultural technology which had not advanced beyond small, slow-moving equipment for ground preparation, planting, and harvesting. Many hedges and other field boundaries are much older than the eighteenth and nineteenth century enclosure movement, especially in western Britain. It has been estimated that there are over 3 million km of field boundary for the 10 million ha of improved farmland in England and Wales. An analysis of conditions in eleven counties by G. M. Locke (1962) gave the following results (Table 15). There was an average of 7 km of field boundary for every km² of improved farmland. Sixty-five per cent was made up of hedges, 15 per cent of materials other than timber, and the remaining 20 per cent wholly or partly of timber. The regional composition of field boundaries varied enormously, with hedgerows becoming progressively more important as one moved south-westwards to reach a peak in Devon.

TABLE 15. TOTAL LENGTH OF FIELD BOUNDARIES IN RURAL AREAS

	km of boundaries per km²	Proportion of total length		
		Hedges (%)	Timber (%)	Other (%)
Ross and Cromarty	0.7	0	74	26
Argyllshire	1.0	1	70	29
Cumberland	6.6	50	28	22
Westmorland	6.9	33	13	54
Yorkshire (N.)	6.8	53	20	27
Yorkshire (E.)	6.5	64	28	8
Warwickshire	7.3	74	18	8
Montgomeryshire	7.5	69	28	3
Hertfordshire	7.5	65	25	10
Essex	6.1	77	16	7
Devon	10.1	86	7	7
Average	7.0	65	20	15

Source: LOCKE, G. M. (1962), A sample survey of field and other boundaries, *Quarterly Journal of Forestry* **56,** 140.

For much of the present century the new agricultural technology that was emerging could be accommodated within the existing pattern of farm sizes and buildings. Tractors were small and slow and could operate economically in small fields. Many of the new advances, such as new varieties of seed, artificial insemination of dairy cattle, and chemical fertilization, could be equally effective on small or large farms, fields, and buildings. In the last 20 years it has been increasingly necessary to replace scarce labour by machinery and enlarge the size of farms and fields. The past decade represented a period of very rapid structural change in Great Britain. The number of farm units declined by 21 per cent between 1960 and 1970, with the total falling from 440,000 to 355,000. The average size of holding rose from 28 ha to 34 ha over the same period. The pace of farm enlargement will probably accelerate during the 1970's and farm numbers could fall to 250,000 by 1980.

Hedgerow removal forms one important part of this structural change. Estimates on the rate of destruction vary enormously. Some authorities have placed it as high as 20 per cent of the hedgerows in England having disappeared over the past two decades. G. R. Allen (1972) estimated that hedgerows were being cleared at the rate of something between 112,000 and 160,000 km each decade since 1950. He considered that this change was a necessary precondition for much of the net output per head in British farming. The most dramatic changes are undoubtedly occurring in the eastern arable counties where mechanized and increasingly specialized cereal production is proceeding apace. A detailed investigation in Huntingdonshire by E. Kenworthy-Teather (1970) showed a 12 per cent decline in total hedge lengths in only 4 years during the second half of the 1960's.

Six main defences may be raised for retaining the hedgerow. It acts as a nature reserve; is a vital component in the ecosystem; functions as a timber reserve, as a shelter belt, and as a good "fence" for livestock; and is undoubtedly a valued aesthetic feature of the English landscape. These advantages are difficult to cost, but arguments for removal have rather clearer economic implications. Land may be gained for farming, having previously been covered by hedges and their associated ditches. Savings can be made since hedges no longer have to be trimmed and ditches to be maintained. Large fields reduce the need for internal roads. Field shapes can be rationalized and the efficiency of mechanized farming increased as awkward corners are removed. Less damage is caused to implements. Vermin and weeds can be better controlled. Blind corners may be eliminated along country lanes. Larger machinery and implements can be used in the new fields, and the way is opened for new techniques such as aerial crop-spraying and seed-sowing. Disadvantages inevitably result. The traditional hedge-bounded rural landscape of parts of eastern England is being replaced by empty, desolate, treeless environments. Ecologists regret the destruction of hedge and bank habitats which acted as wildlife refuges. Some farmers fear that dust storms may take place with the uninterrupted passage of wind over broad, hedgeless expanses of countryside.

Changes in farm buildings often result from farm amalgamation. On the one hand, traditional buildings of visual interest may become redundant and be demolished. It is essential to survey and record such traditional farm buildings to determine which of them might be usefully conserved on amenity grounds. On the other hand, new buildings are being constructed for factory farming, housing machinery, and for other purposes. These buildings are of quite a different scale and style to that normally encountered in the English farming scene. They are usually constructed of standardized materials such as concrete, corrugated iron, or asbestos, and are large enough to form major intrusions in rural landscapes. Efforts need to be made to modify construction plans so that the finished products blend as much as possible into the, albeit modified, rural landscape.

These kinds of structural change are taking place in Great Britain and in other countries which already have modernized farming systems. Serious structural problems are having to be faced in the countries of western Europe where the modernization of farming forms an important aim in economic and social planning. As in eastern Europe, where land reform and collectivization have been implemented, important changes in agricultural patterns are modifying the visual and functional characteristics of rural landscapes.

STRUCTURAL CHANGES IN WESTERN EUROPE

In the late 1940's and 1950's the main agricultural concern of governments in western Europe was to increase the volume of food output. Now many are concerned with trying to cut back production in order to reduce and eventually to avoid surpluses. Social aims, such as raising agricultural incomes, form another reason for governmental intervention. Four types of measures have been introduced to try to achieve this second objective: first, those which support incomes directly; second, those which attempt to improve the efficiency of existing farms; third, those which alter the shape and size of holdings and promote the cession of uneconomic units; and, fourth, those which aid lagging regions by renovating farming and other sectors of regional economies. Forms of intervention have been very complicated, with national governments pursuing their own policies. However, attempts at policy harmonization have been made since the creation of the Common Market.

The results of the first type of intervention have been of limited value. Only slight rises in incomes are normally produced, but the already excessive and inefficient farm population is induced to remain on the land. The second type raises the efficiency of existing farms through grants and loans for land improvements, aiding the formation of co-operatives, and operating advisory schemes. Increased productivity may enable farmers to enjoy a higher standard of living provided that prices of agricultural commodities rise sufficiently. At the present time of agricultural surpluses in western Europe it is arguable from an economic point of view that it is no longer desirable to raise the efficiency of every farmer who applies for assistance.

The third type of intervention involves reorganizing farm structures. If sufficiently radical, this line of attack may offer a more satisfactory solution than the first two types of intervention. Certainly Dr. Mansholt's proposals for the modernization of farming in the Common Market countries by the 1980's laid great stress on the need for structural change to overcome two main problems.

The first stems from the fact that agricultural land in western Europe is divided into small farms, averaging only 12 ha, with two-thirds of them being less than 10 ha in size. Three-quarters of them are too small to permit their labour force to be used rationally. Only 3 per cent are over 50 ha. On average each farmworker in the Common Market managed only 6 ha of land, but his British counterpart managed over three times that amount. The general conclusion was that at least

80 per cent of west European farms were marginal, judged against the technical and economic standards of the late 1960's.

The second structural problem involves the pulverization of individual farms into tiny strips that may be scattered over a large area and are both time-consuming and uneconomic to work as M. Chisholm (1962) has shown. Under such conditions, mechanization may be precluded or might operate only in a highly unsatisfactory fashion. Property fragmentation has several origins. Some examples represent the fossilization of open-field patterns inherited from historic systems of communal farming in areas where an enclosure process has not operated. Others result either from the operation of inheritance laws which demanded an equal division of property between heirs, or from piecemeal reclamation of farmland from the waste or from marshland. In the mid 1950's it was estimated that at least one-half of the farmland in non-Communist Europe was in immediate need of plot consolidation. Conditions varied considerably from country to country. Only 5 per cent of the farmland in Denmark and Sweden was considered to be in need of consolidation, but the proportion rose to 50 per cent in West Germany and Spain and to 60 per cent in Portugal.

PLOT CONSOLIDATION

Consolidation schemes should ensure the amalgamation of scattered plots into compact holdings around farmsteads. Internal divisions on the newly laid out farms should be kept to a minimum necessary for efficient management, the whole ideally being large enough to provide an adequate living for the farmer and his family. Schemes exist in every country of western Europe, but they vary in their scope and approach from simple exchanges of parcels of land to very ambitious programmes of rural management in which consolidation is just one component along with fertilization, building of new or improved roads, water management, and even industrial development, clearance of rural slums, and the construction of new farmhouses. The degree of initiative required for implementing consolidation programmes varies greatly from country to country. In France, for example, plot consolidation can only follow requests from the majority of landowners in the area involved, but in other countries the Government may take the initiative for change in certain designated areas. Once consolidation has taken place it is essential that further pulverization should be avoided, and hence legislation exists to this effect in several countries.

Programmes for plot consolidation operated in France after World War I, but these early schemes were few in number and were concentrated in areas of wartime destruction along the northern frontier. An official policy of *remembrement* was started in 1941. Unofficial exchanges of plots between landowners had, of course, taken place before that date. Fourteen million ha were considered to be in immediate need of reorganization. This represented 40 per cent of French farmland. By 1969 over 6 million ha had been consolidated. Requests for consolidation need to be put forward by three-quarters of local landowners before government-financed schemes may be started. The new consolidated plots need to be of equal productive value to the strips formerly held by each landowner. Hence

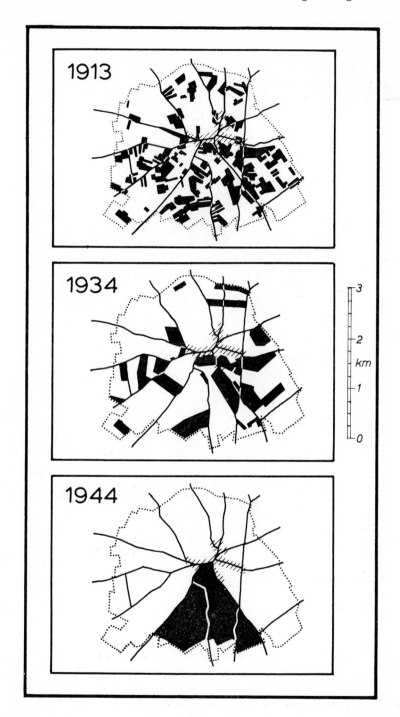

Fig. 7.1. *Remembrement* of Le Bosquet *commune*, Somme *département*, northern France.

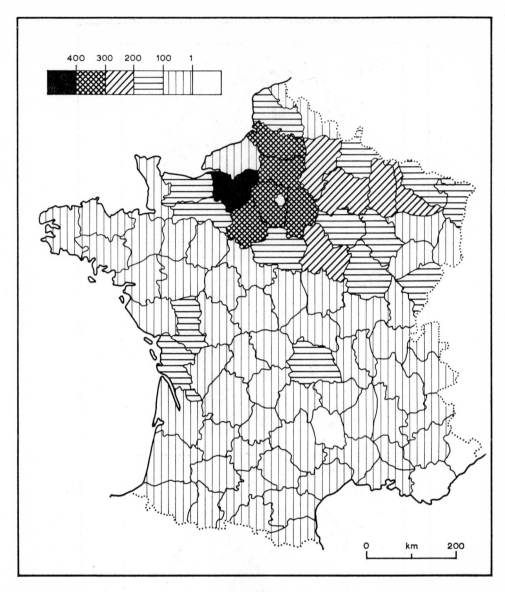

FIG. 7.2. Number of schemes for plot consolidation completed per *département* in
France.

complete consolidation is rarely possible, especially in areas with a diversity of
terrain, soil quality, aspect, and agricultural production. Figure 7.1 records a
rare example of a village where two phases of consolidation have taken place.
The results of the first stage of consolidation, with each landowner still holding a
number of blocks of land, are more typical than those of the second stage, where
large, unbroken blocks of land have been created.

By far the greatest amount of plot consolidation has been achieved in the Paris Basin (Fig. 7.2). Very little has been completed in southern and mountainous areas. Such spatial variations are linked to a range of factors. First, consolidation is a voluntary operation and hence much depends on the degree of conservatism amongst local landowners and their willingness to request for schemes to be started. Second, costs of survey and land reorganization are covered by the Government but a proportion of ancillary costs has to be met by local landowners. Such expenses arise from the need to remove hedgerows and earthen banks, infill sunken tracks, and install new roads and land drains. The total costs of plot consolidation per unit area are much higher in enclosed regions (*bocage*) of western and central France than in the open-field zones of the Paris Basin which already form the most go-ahead and viable agricultural areas in France. Third, far less consolidation has been achieved in areas with small, traditional, owner-occupied farms than in areas with large, often tenant-operated holdings. Such progressive features are found in the northern half of France. Finally, the physical conditions of agricultural production are much better in the Paris Basin and in other lowland areas with rich soils than in the upland zone. Government finances, land surveyors, and agricultural planners have been much more readily forthcoming for the former type of zone than for the latter, where the chance of viable agricultural production in the future is far less sure. In short, areas with the most-favoured environmental conditions for agricultural production and the strongest commercial bias have benefited most from *remembrement*.

Plot consolidation in West Germany operates in a rather different way, being part of either plans for accelerated consolidation or for integrated structural reform. In recent years the advantages of simple acceleration programmes have been stressed by agricultural planners because of their rapidity and much lower cost, by comparison with integral reform. Almost 6 million ha need consolidation for the first time and a further 2.8 million ha that have already been consolidated need a second phase of reorganization in the light of modern requirements. It has been estimated that it will take about 30 years to consolidate land to satisfy the requirements of agriculture at the 1970 level of sophistication. But such requirements will undoubtedly become outdated in the future. Even more radical reorganization will be needed. The slowness of plot consolidation and the constraint which this represents for future farm rationalization characterizes not only France and West Germany but the whole of continental western Europe.

FARM ENLARGEMENT

A further structural problem is centred on the need for farm enlargement. The average size of holding in the Common Market countries was only 12 ha in the late 1960's. Units of such a size were not capable of producing a reasonable income for a family unless involved in very intensive production such as market gardening. Dr. Mansholt urged that farm sizes should be changed radically. This aim would be helped by reducing the direct pressure of population on the land through diverting children of farming families away from agriculture to take on other jobs; encouraging the elderly to leave farming by paying special pensions or annuities;

and by attracting some of the young people in farming to retrain so that they could take on other types of job.

Whilst one may condemn the small size of most farms in western Europe, it is not easy to determine what sizes would be considered desirable for the future. Many countries have aimed to create units for full-time operation ,by two men, thus perpetuating the family scale of farming. Absolute size targets are rarely quoted. Thus Dr. Mansholt's suggestions for changes of farm size were linked to particular aspects of production. He considered, for example, that a suitable farm for the 1980's would have 80–120 ha of cereals, or would raise 40–80 dairy cows, 150–200 head of beef cattle, 450–600 pigs, or 100,000 head of poultry each year. The magnitude of such targets is emphasized when one recalls that in the late 1960's two-thirds of all farms in the Six were under 10 ha in size and an equal proportion of all dairy farmers had fewer than five cows apiece. As D. Warriner (1969) explains: "the optimum size, i.e. the size which maximizes output per man, is a variable dependent on several factors: the density of farm population expressed in the man/land ratio, the type of land use, determined on the one hand by the type of soil and on the other by the market; methods of production determined by the supply of capital and the level of technology. These conditions may vary between countries, and some of them change in the process of development" (p. 38).

Special legislation in some western European countries ensures that when farmland falls vacant it should be used to enlarge neighbouring holdings. But the results of this kind of "natural change" are very slow. It takes a decade to raise the average size of farm in the Common Market by a single hectare. Annuities have been introduced in some countries for payment to elderly farmers who agree to retire from farming and allow their land to be used for enlarging neighbouring farms. Other schemes give financial assistance to younger farmers who elect to retrain for other jobs and allow their farms to be used for restructuring. Special annuities were paid to 123,000 farmers in France during the first 5 years after the scheme was started in mid-1963. Very often only slight increases in farm size were produced. The annuities have been criticized for failing to induce real economic developments and merely providing larger old-age pensions. Similar schemes have operated in Austria, West Germany, the Netherlands, and Sweden.

Agencies operate in some countries to purchase land as it comes freely on to the market and then store it for a while in a land bank. During that time, improvements such as new roads and buildings are implemented and the land is then released for the enlargement of surrounding farms. The French SAFER organizations (*Sociétés d'Aménagement Foncier et d'Établissement Rural*) and the county agricultural boards in Sweden exemplify these kinds of agency which also have powers to regulate ownership changes of any portions of farmland that come on to the market and thereby ensure that average farm sizes continue to be enlarged.

SETTLEMENT REMODELLING

In addition to plot consolidation and farm enlargement, many areas require their settlement patterns to be remodelled so that farmsteads may be sited in the

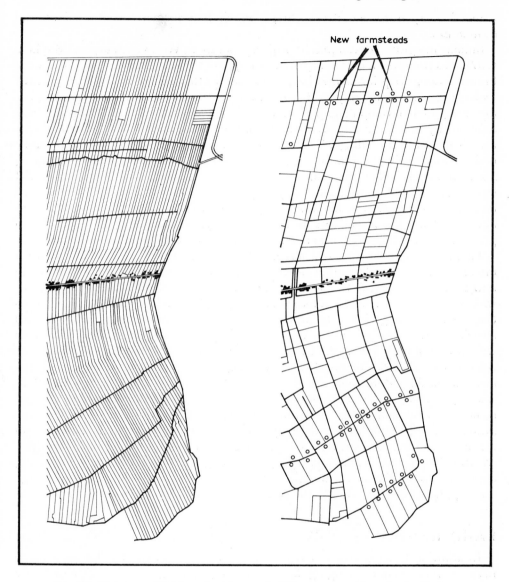

Fig. 7.3. Structure of fields and settlement in Vriezenveen *commune* in the Netherlands before and after integrated land management.

midst of their fields rather than in overcrowded villages or at other points that may be inconvenient for working the land. Policies to achieve this kind of integrated management, involving plot consolidation, removal of rural slums, and construction of new farm buildings, have been implemented in parts of the Netherlands and West Germany. Figure 7.3 shows the result of this type of operation at Vriezenveen in the Netherlands, where fragmented farm strips were regrouped into consolidated holdings and new farmsteads and field roads constructed away from the street village which had formerly contained all the farmhouses.

Similar integrated development projects operate in West Germany. A. Mayhew (1971) has provided a case study in his analysis of the settlement of Mooriem near Bremen. Long-strip farms had been produced in this village through progressive reclamation of low, marshy land in the valley of the Hunte. After the operation of the integrated development project these inconveniently shaped holdings were replaced by square or short-rectangular block farms, and some farmsteads were resettled outside the village. This kind of remodelling process is expensive, since besides constructing farm buildings, public utilities have to be installed, including hard-surfaced roads out to each dispersed farmstead, along with piped water and electricity. It *is*, of course, less costly to provide utilities at farms relatively close to existing settlements rather than at greater distances. But the construction of new farmsteads in such nearby locations is rarely ideal as a long-term solution to problems of land fragmentation and inefficient agricultural production.

Planned settlement dispersal has met other problems of a social as well as a functional nature. It has often proved difficult to encourage farmers to move to new farmhouses outside the old settlements. A. Mayhew (1971) explains that at Mooriem "the problem was a basic one of trying to break the traditional association of the farming family to particular farm buildings and land parcels. This association was strongest in the large farms (owner-occupied), where the farmers were proud of their heritage. It was not until all the overwhelming advantages of resettlement had been demonstrated by small farmers that some of the established farmers considered making the break with tradition" (p. 68). Resettlement in small groups or hamlets is now preferred to a scattering of completely isolated farmhouses. Such a policy cuts the cost of providing utilities and reduces the serious social isolation experienced by families after they have moved away from nucleated settlements. Between 1957 and 1968 3.3 million ha of land were dealt with by integrated development projects throughout West Germany, involving about 6000 resettlements. Unfortunately, the pace was far too slow to make a large and significant contribution to the modernization of West German farming.

LAND REFORM

In addition to such forms of structural change in western Europe there have been movements for changing land ownership through land reform in Italy as well as in the Communist states of eastern Europe. The objectives of land reform in Italy were of a social and political nature, namely to provide farms for landless agricultural workers, thereby creating a "rural democracy", and to enlarge existing smallholdings through the division of massive, under-used estates. Inevitably, land reform also involves important changes in the visual and functional components of the countryside. The end result appears to be just the opposite of farm enlargement and has created further problems for Italian agricultural planners.

Inter-war governments in Italy refused to recognize the existence of agricultural poverty in the south, but political upheavals threatened when soldiers returned

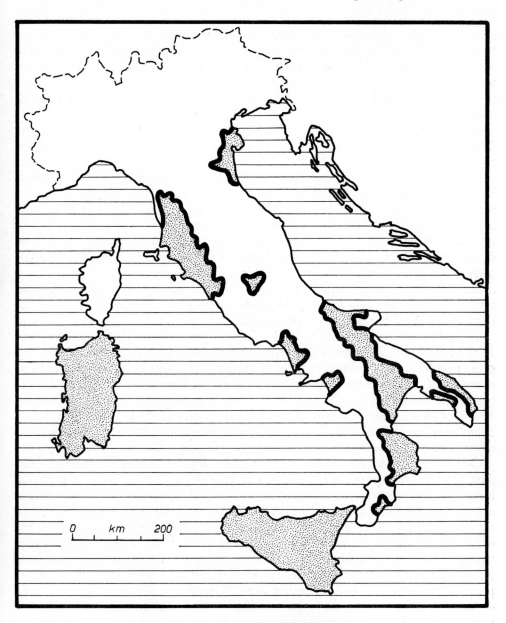

Fɪɢ. 7.4. Land reform areas in Italy.

from military service after World War II. Complicated legislation in 1950 introduced land reform which had the conflicting aims of both modernizing agriculture and providing adequate employment and decent living standards for the agricultural population. Land reform permitted the acquisition and redistribution of land from estates above a 300 ha threshold in the land-reform areas

(Fig. 7.4) comprising 28 per cent of the national territory. The authorities considered that the reform was not necessary in regions where agricultural modernization was already under way. Four-fifths of the land involved in reform was in the south and the islands. D. McEntire (1970) explains that "the land reform was, thus, primarily a Southern program. The nature of the areas selected and the expropriation formula pointed the reform to lands that were either uncultivated or extensively cultivated. Substantial works of land development were necessary to make the expropriated land usable for intensive small-farm cultivation. Thus, the land reform turned out to be in great part a program of land reclamation" (p. 192).

Some 700,000 ha were redistributed. Over 100,000 families, comprising about half a million people, received portions of land between 1951 and 1962. Poorest members of the agricultural community were given priority. Thus 44,500 casual labourers without experience as independent farmers received completely new holdings, with farmhouses located either singly or in small groups away from densely packed agrotowns. In addition, 69,000 parcels of land were redistributed as transfers from renting to ownership or to supplement the holdings of farm labourers who already owned some land but not enough to fully employ the family's labour. The new farms varied in size from 4 ha of irrigated land to 20 ha of poor hill land, with an average of some 7 ha.

The relative success of Italian land reform has varied greatly between regions. In the Metaponto irrigated farms of about 5 ha have been highly successful, but great problems have been encountered on the new farms of other areas such as Sicily and Calabria. Many aspects of the operation have been criticized with the wisdom of hindsight. Costs of installation were high, ranging between $6000 and $9000 for each family. It has been argued that the aim of trying to create a landed democracy in southern Italy in the 1950's was unrealistic and instead at least part of the investment should have been devoted to creating additional industrial jobs.

Other criticisms stem from the small size of many new farms. Units of less than 8 ha have proved quite inadequate to employ a family and provide a sufficient income even in a good year. The physical environment (soil capability, climate, slopes) was not investigated sufficiently before new farms were located and their sizes determined. Many new farms in hilly terrain have now been abandoned, but only after the expenditure of considerable sums on building farmhouses and providing utilities. By 1962 some 15 per cent of all assignees had abandoned their new farms. Many others continued to hold contracts and cultivated the land, whilst the principal breadwinner or some other working members of the family had taken on other forms of employment and sometimes had even migrated to industrial work in the cities of the Mezzogiorno of northern Italy or elsewhere in western Europe.

Inquiries among members of families previously settled on new, dispersed farms as a result of the land reform showed that many considered that living in isolated farmhouses was lonely and unsafe. Instead, many of them returned to their agrotowns from which they continued to cultivate the land as best they could, often returning to a single crop cultivation. D. McEntire (1970) describes the situation in part of Calabria: "Here was a poor, unsanitary village, probably no better in its physical conditions than 20 years before, but crowded with people, while the adjoining countryside was dotted with modern, commodious and empty houses" (p. 198).

Many landless labourers, who became small owner-operators after the reform, were not intellectually prepared to manage their own holdings. The new life proved difficult for many of them, and a vast educational programme was required. But this, too, has been criticized. The reform agencies did almost enerything for the settlers—developing land, laying out farms, building houses, organizing services, constructing settlements, organizing co-operatives, and then introducing selected settlers into this new environment. The assignees "received their farms ready made, without effort on their part, but subject to, for them, some strange and uncomfortable conditions like living on the land, joining a cooperative, and taking directions from a representative of the despised government. Although heavily subsidized, they looked out on 30 years of loan repayment" (McEntire, 1970, p. 198).

Another criticism arises from the fact that the creation of small farms through land reform led to a serious pulverization of property which ran directly counter to what was being attempted by agricultural planners elsewhere in western Europe. All the units were assigned to beneficiaries with 30 year purchase contracts. To prevent speculation, settlers might not alienate their holdings until the total purchase price was paid. New management programmes have been introduced to attempt to counter the small average size of holdings, the continuing fragmentation of farm units, and the general weakness of institutions for the diffusion of technical innovations and really modern systems of marketing.

LESS LAND IN AGRICULTURAL USE

A final aspect of rural restructuring involves the prospect of major changes in land use in the future. Dr. Mansholt argued that, whilst modifications in commodity prices, improvements in agricultural marketing, and drastic changes in farm sizes might help reduce or even avoid surpluses of farm products, a more rigorous change was required if farming was to be converted from a tradition-bound way of life to a modern business activity. This, he considered, would involve a reduction of the total farmed surface in the Common Market by 7 per cent before 1980. Dr. Mansholt suggested that four-fifths of the liberated agricultural land should be used for forestry. The implications of this idea will be considered in Chapter 8. The remainder of the land thus released would be converted to national parks and other types of recreation area. In fact, such proposals represent an accelerated extrapolation of existing trends involving the conversion of poor farmland to other uses or its falling into disuse.

Urbanites are certainly in search of more recreation space and, as the productivity of farming continues to rise, a smaller amount of land is required to produce a given quantity of food, as Swedish data clearly illustrate (Fig. 7.5). Nevertheless, the two issues are not as directly complementary as one might suppose at first. Visitors to the countryside do not wish to see uncultivated, scrub-covered landscapes that would surely result once farmers and their livestock were

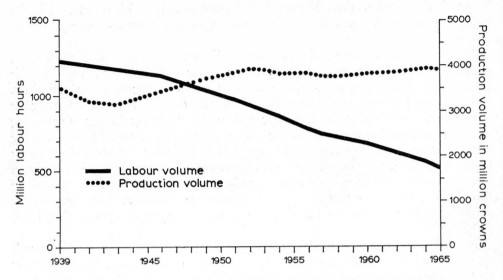

FIG. 7.5. Changes in labour inputs and production volume of Swedish agriculture, 1939–65.

removed from their traditional "manicuring" role. In addition, farming still employs 17 per cent of the workforce in the Common Market countries. It is true that numbers are decreasing rapidly, but it is unlikely that the total agricultural population in the Six could be reduced to 5 millions in 1980 as Dr. Mansholt proposed. Such a figure would represent only a half of the 1970 total and one-quarter of that for 1950. One can envisage a situation in the future when a proportion of farmers might well be in receipt of financial assistance not to help them raise productivity but to encourage them to act as "park keepers" for stretches of country deemed to be primarily of visual and recreational importance.

REFERENCES AND FURTHER READING

Structural changes in both western and eastern Europe are considered briefly in:

CLOUT H. D. (1971) *Agriculture*, Macmillan Studies in Contemporary Europe, London.

Changes in farm numbers and the micro-structures of British farming are considered in:

ALLEN, G. R. (1972) The scope of rural life: the influence of agriculture in the 1970's. Mimeographed paper for the Town and Country Planning Association's Conference in London.
ASHTON, J. and CRACKNELL, B. E. (1961) Agricultural holdings and farm business structure in England and Wales, *Journal of Agricultural Economics* **14,** 472–506.
BAIRD, W. W. and TARRANT, J. R. (1972) Vanishing hedgerows, *Geographical Magazine* **44,** 545–51.
KENWORTHY-TEATHER, E. (1970) The hedgerow: an analysis of a changing landscape feature, *Geography* **55,** 146–55.
LOCKE, G. M. (1962) A sample survey of field and other boundaries, *Quarterly Journal of Forestry* **56,** 137–44.
WELLER, J. (1967) *Modern Agriculture and Rural Planning*, Architectural Press, London.

Plot consolidation and farm enlargement schemes in western Europe are discussed in:

BAKER, A. R. H. (1961) Le Remembrement rural en France, *Geography* **46**, 60–62.

CHISHOLM, M. (1962) *Rural Settlement and Land Use*, Hutchinson, London.

CLOUT, H. D. (1968) Planned and unplanned changes in French farm structures, *Geography* **53,** 311–15.

FRANKLIN, S. H. (1969) *The European Peasantry*, Methuen, London.

LAMBERT, A. (1963) Farm consolidation in western Europe, *Geography* **48,** 31–48.

MAYHEW, A. (1970) Structural reform and the future of West German agriculture, *Geographical Review* **60,** 54–68.

McENTIRE, D. and AGOSTINI, D. (eds.) (1970) *Towards Modern Land Policies*, University of Padua Press, Padua.

OECD (1965) *Obstacles to Shifts in the Use of Land*, Paris.

PERRY, P. J. (1969) The structural revolution in French agriculture, *Revue de Géographie de Montréal* **23,** 137–51.

Examples of integrated development projects are found in:

LAMBERT, A. (1961) Farm consolidation and improvement in the Netherlands, an example from the Land Van Maas en Waal, *Economic Geography* **37,** 115–23.

MAYHEW, A. (1971) Agrarian reform in West Germany: an assessment of the integrated development project Mooriem, *Transactions Institute of British Geographers* **52,** 61–76.

Discussions of land reform and structural changes in Italy are found in:

DICKENSON, R. E. (1954) Land reform in southern Italy, *Economic Geography* **30,** 157–76.

FRANKLIN, S. H. (1961) Social structure and land reform in southern Italy, *Sociological Review* **9,** 323–49.

KING, R. (1970a) Structural and geographical problems of south Italian agriculture, *Norsk Geografisk Tidsskrift* **24,** 85–97.

KING, R. (1970b) Land reform in Apulia–Lucania–Molise, *Norsk Geografisk Tidsskrift* **24,** 149–59.

KING, R. (1971a) Land reform: some general and theoretical considerations, *Norsk Geografisk Tidsskrift* **25,** 85–97.

KING, R. (1971b) Italian land reform: critique, effects, evaluation, *Tijdschrift voor Economische en Sociale Géografie*, **62,** 368–82.

McNEE, R. (1955) Rural development in the Italian South, *Annals Association of American Geographers* **45,** 127–51.

WARRINER, D. (1969) *Land Reform in Principle and Practice*, Oxford University Press, London.

Reductions in the area devoted to agricultural uses in the USA are discussed by:

CLAWSON, M. (1971) *America's Land and its Uses*, Resources for the Future Inc., Johns Hopkins Press, Baltimore.

HART, J. F. (1968) Land abandonment of cleared farmland in the eastern USA, *Annals Association of American Geographers* **58,** 417–40.

CHAPTER 8

FORESTRY AS A USER OF RURAL LAND

SINCE the beginning of the century the landscapes of many rural areas in the developed world have been changed dramatically by the intrusion of woodland. This is particularly true in upland regions and other areas where agricultural production is relatively impoverished. In severely depopulated regions, such as parts of central France, arable land has fallen out of cultivation, and broad areas of communal pasture are no longer grazed. Afforestation has taken place in such regions at a variety of scales, ranging from plantations covering hundreds of hectares of village-owned land to tiny "postage-stamp" plantations on privately owned property, often held by absentee urban landlords with no interest in working the land or leasing it to neighbouring farmers. The second scale of afforestation can be particularly serious in areas where property ownership is fragmented. "Postage-stamp" plantations are frequently much too small to be managed for the rational and viable production of timber. They severely hinder local schemes for farm enlargement or property consolidation. One of the major problems faced by forestry managers in Continental Europe is how to encourage private landowners to group together so that large blocks of co-operative woodland may be planted which will be of sufficient dimensions for the rational production of timber in the future.

As agricultural productivity continues to rise it is clear that the total area devoted to intensive farming in the developed countries will have to be cut back if costly surpluses are to be avoided. This assumes, of course, that suitable channels for distributing all food surpluses to the hungry nations of the Third World will not be found. In 1968 Dr. Sicco Mansholt reported that large sections of less productive farmland in the Common Market countries should be withdrawn from agricultural use. He recommended cutting back the total farmed area of 71 million ha by 5 million ha before 1980. One fifth of the land released from agriculture might be devoted to national parks and other forms of recreational area. The remainder should be used for timber production. Figure 8.1 summarizes the land capability in the Six and depicts the impoverished agricultural regions where major land-use changes may be expected in the future. The Vedel report on agricultural rationalization suggested that 5 million ha might be withdrawn from farming use in France alone. Similar policies for reducing the total farmed area are already operational in other parts of the world, for example in Sweden and the United States.

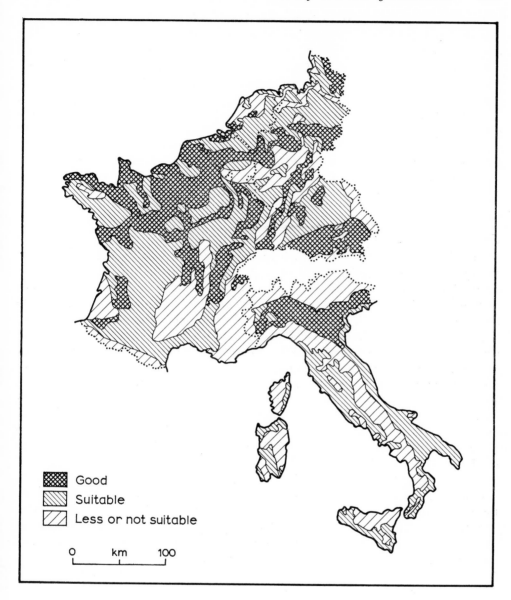

Fig. 8.1. Land capability of the six Common Market countries for agricultural production.

It would seem unlikely that west European agriculture can be managed in such a way as to reduce farmland surfaces at the rates proposed by Mansholt or Vedel. Nevertheless, the tendency for arable land and improved pasture to give way to scrub and timber already characterizes most upland areas and will undoubtedly continue to do so even more strongly in the future. Demands for softwood timber for pulp and paper production rise each year, reflecting the requirements of the consumer society, and have to be met very largely from costly imports. In Great Britain, for example, over 90 per cent of timber supplies are imported at an annual cost of more than £500 million. Current trends of agricultural decline in the uplands and increasing requirements for softwood combine with the planners' proposals and suggest that the rural landscapes of less productive agricultural regions will contain a significantly greater tree cover in the future than at present.

Techniques and agencies for forest management, together with rates of afforestation, vary from country to country. The following discussion will concentrate on activities in Great Britain, especially with reference to the work of the Forestry Commission. However, many of the issues which will emerge from such a restricted discussion have much broader application and have also been experienced in many other countries.

AFFORESTATION IN GREAT BRITAIN

At the beginning of the twentieth century the total woodland area in Great Britain covered 1.13 million ha, some 5 per cent of the land surface. As early as 1909 a Royal Commission on coastal erosion drew attention to the condition of Britain's woodlands. It proposed to convert comparatively unprofitable land into forests and hoped thereby to stem the tide of rural depopulation in some upland areas. But in the next decade 182,000 ha of timber were felled to help meet the country's needs during World War I. The inevitable reduction of Britain's forest resources showed both the importance of timber in the modern industrial economy and also the need for large domestic supplies of wood for use in emergencies. In 1916 a forestry sub-committee had been appointed to consider the best means of conserving and developing the nation's forest resources. Two years later it recommended that 710,000 ha should be afforested so that "large areas of the United Kingdom, now almost waste . . . might be put to their best economic use" (quoted in Matthews *et al.*, 1972, p. 26).

The Forestry Commission was thus founded in 1919 and was charged to create new forests, encourage private forestry, and help maintain an efficient timber trade. In the inter-war period the Commission planted 149,000 ha of trees in Great Britain. Nevertheless, an initial shortage of funds restricted its volume of planting which did not manage to compensate for wartime losses. However, this target was, in fact, achieved, since 81,000 ha were planted by private landowners. More felling of mature and semi-mature timber took place during World War II. This inevitable action reinforced arguments for more rapid afforestation in order to maintain a large reserve of timber in Britain.

In 1943, even before hostilities had ended, the Forestry Commission reported on the problems of post-war afforestation and suggested that by A.D. 2000 there should be 2 million ha of properly managed woodland in Great Britain. Such an area was considered to be "required for national safety and will also provide a reasonable insurance against future stringencies in world supplies. There are also valuable contingent advantages associated with forests, such as the development of rural Britain. . . . The 2 million ha should not merely be planted with trees but also systematically managed and developed. This conception entails, among

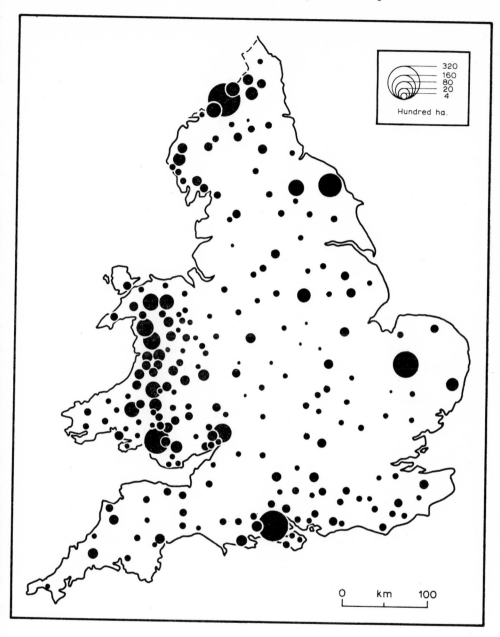

Fig. 8.2. Forestry Commission land in England and Wales, 1966.

other things, the continuous application of good silviculture, the development of markets and internal transport, and the settlement of forest workers in a good environment."

Two-fifths of the 2 million ha target for the end of the century would be achieved by replanting old woodland areas. The greater part would involve planting of new land. If this programme were to be achieved, Great Britain would have 8.8 per cent of its total land area under timber. Even so, it would still be one of the least wooded countries in Europe in spite of the fact that climate and soil conditions in Great Britain are generally well suited to tree production. The average annual growth of properly managed trees in Great Britain is nearly twice that recorded in eastern and central Europe.

The combined area planted annually by the Forestry Commission and by private landowners increased from 1620 ha to 16,200 ha in the 20 years inter-war period. The annual planting rate fell to 4000 ha during World War II, but when hostilities ceased it rose again to reach a record 40,000 ha in 1961. Since then the annual rate has remained at about 36,000 ha. Threequarters of this total involves extending the wooded area. Restocking cleared woodland accounts for the remainder. The Forestry Commission has acquired areas of Royal forestland and has purchased or leased tracts of felled woodland, moor, and mountain from private owners. Much property has been acquired from large landowners keen to sell land in order to obtain capital or to settle death duties. The Commission has often taken over marginal land in upland areas previously used for sheep farming. Figure 8.2 shows that most Forestry Commission land in England and Wales is found in upland areas with the exception of the New Forest in Hampshire and Thetford Chase in East Anglia.

Nine-tenths of the current industrial and commercial demand for timber in Great Britain is for softwoods. Most planting by the Commission is designed to help meet such demands in the future. In any case, much of the available land for planting or replanting is of poor quality and would not satisfactorily produce hardwoods. Such an emphasis on planting conifers by the Forestry Commission aroused considerable criticism from vocal sections of the public on the grounds that conifers were alien to British landscapes and that amenity was being sacrificed when they were planted. However, recent public-opinion surveys have shown that visitors to British forests do not dislike conifers to the degree that was previously supposed. In fact the Commission also planted over 100 million broadleaved trees in the past two decades. Great care is now taken to avoid ruler-straight edges to newly planted woodland blocks as part of the Commission's growing concern with landscaping its forest domain.

Timber-planting by private landowners increased dramatically from 2000 ha each year between 1920 and 1946 to an average of 15,000 ha in the 1960's. This trend has been encouraged by the availability of special tax concessions for afforestation. By 1966 a total area of 1.38 million ha was afforested, covering 7.5 per cent of the land surface of Great Britain (Fig. 8.3). The ownership of British woodland has changed fundamentally since the creation of the Forestry Commission. In 1914 97 per cent of all woodland had been privately owned, but now the proportion has fallen to two-thirds of a much larger total area.

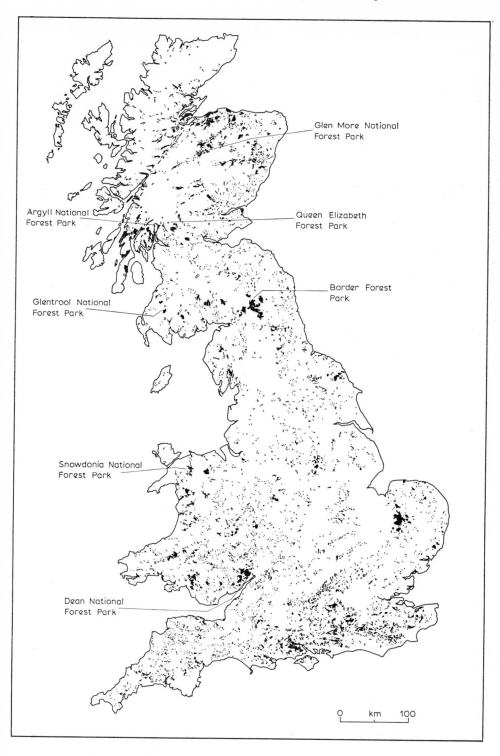

Fig. 8.3. Woodland and national forest parks in Great Britain, 1966.

In spite of important achievements, the Commission has been restricted in its activities by its inability to acquire sufficient plantable terrain in upland areas. This has been due to six main reasons. First, the Commission needs to obtain approval from the Ministry of Agriculture (or the Secretary of State for Scotland) before farmland can be taken for forestry. This amounts virtually to a prohibition regarding the better grazing land. Second, the relative prosperity of agriculture in post-war Britain made it possible for landlords to retain poor quality land even where it was not being used fully. Third, landlords are reluctant to disturb tenants by selling or leasing unproductive parts of occupied holdings. In any case, they have statutory protection for security of tenure. Fourth, competing claims exist in upland areas, such as setting land aside for military training or as water catchments, as well as for afforestation. Fifth, forestry has to be viewed as a long-term investment rather than one which will yield a quick return. In the words of the Earl of Bradford: "most of us operate in hope for our grandsons" (Donaldsons, 1969, p. 86). Finally, as the Donaldsons have shown, "when all is said, the great underlying reason for the failure of the Forestry Commission to acquire land seems to be that the State policy for forestry has been at loggerheads with the State policy for agriculture. . . . Forestry has been sponsored by the State and agriculture has been subsidized by the State, and different men in different offices have worked hard for the advantage of each" (p. 88).

The existing pattern of piecemeal afforestation by both the Forestry Commission and by private landowners is considered unsatisfactory by some foresters because the use of land, labour, and capital is widely and expensively dispersed. Proposals have been made by J. D. Matthews (1972) and others for designating "forestry development areas" in which afforestation would be concentrated to meet modern industrial and social requirements. If they were to be established, forestry development areas would stimulate economic activity in remote regions, thereby providing employment and broadening the range of local skills. They would produce a sustained supply of timber for a variety of wood-using industries which, hopefully, would be installed within their boundaries. Soil fertility would be maintained or improved in forestry development areas. Water supplies could be protected and the aesthetic value of the landscape possibly increased. Ancillary activities, such as sport and recreation, would be encouraged in development areas where the existing pattern of land ownership would, however, remain unchanged.

Special incentives might be provided if forestry development areas were to be designated. These might include constructing new roads for timber transportation; providing guaranteed fiscal relief on the capital cost of machinery and equipment for developing forestry and forest industries; installing electric power, piped water, housing, and other essential services at competitive rates; and even building advance factories or allocating comparable financial assistance for wood-using industries.

Forestry development areas would be most appropriate in areas where the following conditions could be satisfied. Site and climate would need to be favourable for the rapid growth of high-yielding coniferous trees. Each planted area would cover at least 12,000 ha and preferably over 20,000 ha. Suitable sites would be needed for installing integrated wood-using industries such as sawmills and works for making pulp-board and paper. These works would require sufficient water, cheap energy, and good transport links to markets. A suitable small town or large village would be required for further development as a "key" settlement in each area. Forestry management could be integrated with farming, sport, and recreation in each development area, but timber would form the main occupier of land. The planting programme would be sustained for a sufficient period to permit the adjustment of housing, schools, and other social facilities and thereby create a permanent force of skilled labour for forestry.

BROADENING AIMS OF FOREST MANAGEMENT

The Forestry Act (1967) consolidated previous enactments between 1919 and 1963 and charged the Forestry Commission with the general duty of promoting the interests of forestry; the development of afforestation; the production and supply of timber and other forest products; and the establishment and maintenance in Britain of an adequate resource of growing trees. Nevertheless, modern forestry policies recognize the need for closely integrating forestry with agriculture and other forms of land use. Thus, in recent years, the aims of the Forestry Commission have broadened from extending and managing the woodland domain for the economic production of timber to include a variety of other objectives. These were enunciated after the 1963 review of the Commission's programme. They included: (a) to provide employment in rural areas, especially those affected by depopulation, thereby maintaining a skilled labour force; (b) to give due attention to the aesthetic and protective roles of the forest and to encourage open-air recreation; (c) to foster industrial and social development ancillary to forestry; and (d) to encourage the orderly development of private forestry. A further review in 1971 placed greater emphasis on the potential role of Forestry Commission land as recreational space in Britain.

Agriculture can benefit directly from its juxtaposition with forestry since shelter is provided for livestock and grass matures earlier in the lee of plantations. The water-supply industry also benefits, since run off is slowed down once a tree canopy has been formed and the silting of reservoirs is thereby reduced. Forestry diversifies rural employment facilities and can help stabilize or even increase the local population. It is difficult to calculate the number of people employed directly in planting, tending, and harvesting forests. Ratios of 1 man per 40 ha are applicable to fully productive forests where planting is balanced by the felling of mature trees. However, the productivity of woodmen involved in planting has risen in recent years, and British forests are still young and therefore producing below their potential yield of wood. The current ratio for Britain as a whole is perhaps nearer 1 man per 56 ha. There are, of course, great local variations depending on the following factors. The size of the total forest area and the relative proportions actually planted or awaiting plantation will be important. Large plantations require fewer workers per unit area than small, scattered woodland blocks. Local differences in the duration of planting programmes, types of terrain, and the degree of modernization of ground preparation and planting will affect labour inputs. These will depend also on the volume of timber to be cut, whether it is to be sold standing to a merchant or to be harvested by the forest staff, and whether it is to be processed at a local sawmill or moved beyond the forest.

The use of generalized ratios can lead to inaccuracies in estimating the local significance of forestry as a source of jobs in rural areas. Local employment increased five or six times following the changeover from extensive pastoral farming to forestry in parts of north-west Scotland. But in central Wales, for example, with a different agricultural system, there has been little change in the number of men employed per 100 ha following the change from farming to forestry. The contribution of forestry to employment is not only in terms of growing wood but also in processing it. The generally accepted ratio between these two sectors of the industry is in the order of 1 : 4 or 1 : 5. If wood-using industries are sited away from the producer forest, much of the potential benefit for increasing employment in rural areas is lost. By contrast the greatest gain is obtained when an integrated industry involving both primary (pulp-mill, board-mill, or sawmill) and secondary processing (paper or container making) is located close to the producer forests.

The spread of afforestation has served to retain population in some upland areas which were being seriously depopulated. At first the Forestry Commission established dispersed holdings and houses for forestry workers in newly afforested areas. Since 1945 it has adopted a policy of concentration and has established a number of forestry villages to house workers and their families in major planting areas. Rural agglomerations have been set up with adequate population at hand to deal with forest fires. Workers have only short journeys to work in the forest, and are able to make use of the shops and social facilities that have been installed in the new settlements.

After several decades, forest villages have been praised for their ability to retain a stable labour force in afforested areas. But criticisms have also arisen, both from within and beyond the Forestry Commission. G. B. Ryle (1969) records that "there appeared in the heart of several of the larger forests groups and terraces of new houses. . . . Each house was good and well planned with up to date amenities for townsfolk. Each house was just like its neighbour. Each tenant had the same job and the same wage packet and the same boss as his neighbour" (p. 81).

Opinions have been mixed regarding the social success of forest villages. Numerous problems were encountered in the early years after their establishment for the following reasons. Forest planners were attempting to create communities from collections of strangers who appeared to lack common social interests. There was an undesirable lack of social diversification in these settlements since all the inhabitants were of one "class". In addition, forestry workers were living in tied houses, owned by the Forestry Commission, and often felt ill at ease in them. This was heightened by the fact that many of the forest workers and their families were urban in origin and did not take readily to life in the countryside. In the Anglo-Scottish borderlands, for example, where eight forest villages were established, there was a considerable turnover of population with many folk leaving through dissatisfaction. Forest villages had been planned to contain a range of social facilities (shops, a pub, a church, a community hall), but initial populations were too small to support these facilities and they were not therefore introduced immediately. Population drifted away because of inadequacies in the social environment, amongst other reasons. Nevertheless, in spite of such teething troubles many of the forest villages have proved successful, and a strong community spirit has developed in them, especially in recent years.

Such a policy of establishing new settlements may be essential in previously uninhabited areas. But declining or static villages and small towns might be strengthened through the addition of forestry workers and their families in most of the remoter parts of Britain where afforestation is taking place. A decision along these lines was taken for Wales in 1954. New houses for Forestry Commission workers have been sited where they help to infuse new life into the schools, shops, and community facilities of existing villages. This policy to bolster settlements and to diversify, as well as to enlarge, their employment base, has now become more widespread.

FORESTRY AND RECREATION

As early as 1931 the Forestry Commission decided to open some of its land to the public in areas designated as National Forest Parks. These are areas of exceptional beauty made available for public access and enjoyment so far as the requirements of timber production permit. This concept of the national forest park preceded the designation of national parks in England and Wales by many years, and showed an early recognition of amenity issues by the Commission. The first national forest park was established on the banks of Loch Lomond in Argyllshire in 1936. Other parks were set up in Glentrool, Glenmore, Dean Forest, Snowdonia, and Hardknott in the Lake District (see Fig. 8.3). The final park has since been abandoned but two others have been designated, namely the Queen Elizabeth Forest Park in the Trossachs (1951) and the Border Forest Park (1955).

Forest parks cover a total of 175,000 ha in Great Britain and attract over 350,000 campers each year. In addition to being provided with camp sites they contain forest trails, information centres, and car parks, the latter mainly on forest perimeters. Resident wardens supervise camp sites where charges are made, but there are no entrance fees to the State-owned forests. The parks provide a variety of recreation facilities from fishing, sailing, and shooting, to skiing.

The Forestry Commission is becoming increasingly aware of the recreational value of its estates, where over 15 million day visits were recorded in 1969. Most forest parks are located in Scotland and northern England, and their full potential is only realized in the summer months when long-stay visitors spend their annual holidays in such regions. The Queen Elizabeth Park is an exception since it provides all-the-year recreation space for the citizens of Glasgow and Edinburgh. Southern England and the Midlands lack forest parks, and in these regions both private and Forestry Commission woodlands are experiencing increasing pressure from day trippers. H. E. Bracey (1970) has posed the question "Perhaps we need some new forest parks planted now for the next generation in the national amenity interest in the first place and its economic interest in the second. And more of them in England even if they are less profitable financially" (p. 208).

In 1958 the objectives of the Forestry Commission were broadened for creating a large, standing reserve of timber for use in a war of attrition, to managing and extending the national forest estate. Ministerial announcements in 1958 and 1963 and the Commission's annual report for 1970/1 stressed that part of this added function involved catering for recreation. Public access had certainly been permitted to forest parks and to some other forests prior to that time, but the attitude of the Forestry Commission was ambivalent because of the fire hazard. This was particularly severe in areas with young stands of inflammable conifers. Recently the Commission has declared its intention to designate "open forests" where the public would be welcomed and fairly free pedestrian access would be permitted. These "open forests" would be provided with carefully sited car-parks, simple picnic stops, and some short-stay camping sites.

The Commission has sponsored surveys of public recreation in Cannock Forest, Allerston, Loch Lomond, and Glen More. These forests are located at varying distances from major concentrations of population, with Cannock Forest only 27 km from the centre of Birmingham. The recreational use of the four forests varied in detail, but the reports showed "only 6 per cent of use was concerned with nature studies, including ornithology. It seems to be easy to over-estimate the interest in nature studies among forest visitors. In general, the visitors' desire is to escape from cities and from their fellow men, to find a picnic place for the family and to have a country walk. . . . There is an obvious risk that the interests of the forest manager, which commonly may lean towards natural history, tend to be magnified in the assumed demands and interests of the public users" (Mutch, 1968, p. 83).

Over 90 per cent of visitors came to the forests by car, and clearly the provision of more parking facilities either within or on the margins of forestland is essential. Two-thirds of the visitors questioned indicated that they wanted improved access facilities within the forests. Signposted footpaths were requested most frequently, followed by forest maps, better-surfaced and wider roads, and more parking places. Facility improvements were also requested. These included—in decreasing order of significance—sanitation, picnic tables, water points, restaurants and garages, shelter belts, and information offices.

Visitors made it clear that they deplored "public park" conditions, and sought uncrowded forests for their pleasure. Few visitors saw anything in timber production that was inherently antagonistic to their recreation, and they implied thereby that multiple use of land was quite feasible for timber-growing and for recreation. Most visitors reported that they liked conifers. This was a surprising result bearing in mind the strong opposition which had been expressed previously from some vocal quarters. In fact, 91 per cent of the visitors questioned thought that conifers provided a suitable environment for recreation. Only 1 per cent found them unsatisfactory and the remainder did not know. Two-thirds of visitors (67 per cent) said that they found conifers more attractive than hardwood trees. Thirteen per cent found them less attractive, 17 per cent equally attractive, and 3 per cent had no opinion. Even on their first visits, very few respondents thought evergreen conifers made the forest unsuitable for recreation. Indeed, there was no significant difference in attitude to conifer forests between respondents on their first visit and those on second and subsequent ones.

It is clear that there is already a brisk use of national forests in Britain for recreation activities, and the demand is increasing rapidly. A rate as high as 45 per cent per annum has been mentioned for some forests. Most visitors seek seclusion, quietness, and freedom to wander afoot. Most are in favour of car-free areas. However, the planning of access for cars, while preserving traffic-free zones and overall quietness, will be quite essential if the forests are to provide a service to the community commensurate with the heavy capital investment in them. Satisfactory recreation facilities can be provided in evergreen forests that were originally planted for timber production. Diversity of age is likely to enhance recreational and aesthetic values even more than diversity of species. One thing is certain and that is that far greater attention will have to be paid to the visual implications of land-use change than has been customary in the past.

The Forestry Commission's annual report for 1970/1 contained a critically important statement on the role of recreation in forestry management in Great Britain. This report outlined management practices that will greatly diversify the use of some sections of British woodland. The Commission stated that in the future its policy was to develop the unique recreational features of its forests, particularly where they are readily accessible to large numbers of visitors from cities and holiday centres. This would be done in conformity with the Commission's statutory powers and obligations, within the financial resources available and subject to the primary objective of timber production. The Commission emphasized that its programme for recreational development would neither injure the forest environment nor conflict with its conservation. In order to achieve these aims the Commission will prepare a series of recreation plans in consultation with local planning authorities and other interested bodies such as the Nature Conservancy, the Countryside Commission for England and Wales and for Scotland, the Sports Council, and the Commission's own regional advisory committees which advise on amenity and recreational matters.

The Commission recognized that in the immediate future nearly all its forests are likely to become places where people will want at least to walk and picnic, if not to indulge in other forms of recreation. Its policy is to allow the public to enter on foot all its forests except those subject to agreements which would be infringed by unrestricted access, provided that this access is not in conflict with the management and protection of the forest and is subject to the Commission's bylaws. Access by the public for air and exercise, including the use of forest walks and picnic places, will be free of charge. The Commission will provide facilities for the needs of visitors, including small car-parks and information centres. These will not be expected to be completely self-supporting, and charges will be made wherever feasible. Larger facilities will also be provided, such as sites for camping and caravanning plus some relatively large car-parks and information centres. Charges will be made for their use. The regular use of forest roads by cars for recreational purposes will be prohibited except where necessary for access to camping and picnic places and to car-parks in designated places. The existing status of the forest parks will be maintained. Those forests and parts of forests where access and recreation are particularly encouraged will be clearly signposted.

The Commission's forests provide opportunities for many types of leisure pursuits such as fishing, pony trekking, sailing, and nature study. The Commission's policy is to continue to manage facilities for field sports in accordance with accepted codes of practice. Priority will, however, be given to recreational activities by the general public wherever any conflicts of interest are likely to arise. Opportunities for fishing and horse-riding will be extended. The Commission seeks to obtain sufficient revenues from special activities to cover costs incurred. Such proposals represent a radical reappraisal of forestry policy in Britain, and it is clear that an extremely important new phase of woodland use is about to come into operation in the immediate future. Just how the policy statement will be implemented, which forests and parts of forests will be affected, and which activities will be authorized in particular areas, can only be matters for conjecture at this stage.

REFERENCES AND FURTHER READING

Aspects of afforestation in Great Britain are discussed in:

BRACEY, H. E. (1970) *People and the Countryside*, Routledge & Kegan Paul, London.
CAMPBELL, J. (1970) Britain's third forest, in *The Changing Uplands*, Proceedings and papers of a conference sponsored by the North Pennines Rural Development Board and the Country Landowners Association at Harrogate, pp. 35–40.
DONALDSON, J. G. S. and DONALDSON, F. (1969) *Farming in Britain Today*, Allen Lane; the Penguin Press, London.
FORESTRY COMMISSION (1971) *Fifty-first Annual Report, 1970–71*, HMSO, London.
GRAYSON, A. J. (1967) Forestry in Britain, in ASHTON, J. and ROGERS, S. J. (eds.), *Economic Change and Agriculture*, Oliver & Boyd, Edinburgh, pp. 168–89.
HOUSE, J. W. (1956) Afforestation in Britain: the Anglo-Scottish borderlands, *Tijdschrift voor Economische en Sociale Geografie* **47**, 265–76.
MATHER, A. S. (1971) Problems of afforestation in north Scotland, *Transactions Institute of British Geographers* **54**, 19–32.
MATTHEWS, J. D. *et al.* (1972) Forestry and forest industries, in: ASHTON, J. and HARWOOD LONG, W. (eds.), *The Remoter Rural Areas of Britain*, Oliver & Boyd, Edinburgh, pp. 25–49.
MUTCH, W. E. S. (1968) Public recreation in national forests: a factual survey, *Forestry Commission Booklet 21*, HMSO, London.
RYLE, G. B. (1969) *Forestry Service*, David & Charles, Newton Abbot.
SHARP, T. (1955–6) Forestry villages in Northumberland, *Town Planning Review* **26**, 165–70.
VERNEY, R. B. (1970) The Commission's forests, in *The Changing Uplands*, Proceedings and papers of a conference sponsored by the North Pennines Rural Development Board and the Country Landowners Association at Harrogate, pp. 27–33.

Problems associated with postage-stamp afforestation are considered in:

CLOUT, H. D. (1969) Problems of rural planning in the Auvergne, *Planning Outlook* **6**, 29–37.

CHAPTER 9

LANDSCAPE EVALUATION

LANDSCAPE forms a valuable but vulnerable resource, especially in densely populated countries such as England and Wales where 15,000 ha of countryside are abstracted from "rural" uses each year for various forms of "development". Formidable landscape changes result. Planners are in the difficult position of having to establish a checklist of landscape qualities that will help them decide which proposals for development they will sanction and which they will refuse on the grounds of landscape "erosion". Particular attention needs to be paid to the future of "beautiful" landscapes. In the words of R. J. S. Hookway and J. Davidson (1970): "some identification of the most valued areas must be made and adequate provision ensured for their protection or re-creation" (p. 18). It is true that national parks and "areas of outstanding natural beauty" have been defined, but such designations were achieved without recourse to any recognized methodology for the evaluation of rural resources. What is now needed is a technique for classifying landscapes so that appropriate planning action may be taken for their future management. A. E. Weddle (1969) argues that methods of landscape evaluation "should give a perception that is sharper than that of the casual observer, and a clear understanding of the general response likely to be aroused by landscape qualities, or indeed by their loss" (p. 387). S. B. K. Clark (1968) has suggested that the following range of landscape types might usefully be identified: (i) landscapes with special characteristics to be preserved in their existing conditions at all costs; (ii) beautiful areas where development should be very carefully controlled; (iii) monotonous agricultural areas; and (iv) other areas with no special landscape interest.

Experimental studies have been undertaken in recent years in order to evaluate rural resources in various parts of Britain and abroad. Clearly such an evaluation operation involves more than simply identifying and mapping land-use variations. But such an approach is a fundamental contribution to the total investigation. The second part of the operation has been explained by S. B. K. Clark who stressed that the values that people attach to landscapes and to the countryside in general must be sampled in an attempt to arrive at the "mean of people's comparative reactions" for given areas of terrain. In other words, "whichever way it is looked at, landscape can be classified from a human as well as a purely factual aspect" (p. 19). This is where the real complications are encountered since each individual has his own subjective perception and appraisal of different landscapes. D. Lowenthal (1967) has explained: "in daily practice, we all subordinate reality to the world we perceive, experience, and act in. We respond to and affect the environment not directly, but through the medium of a personally apprehended *milieu*. This *milieu* differs for each of us according to his personal history: and for each of us it varies also with mood, with purpose, and with attentiveness. What we see, what we study, and the way we shape and build in the landscape is selected and structured for each of us by custom, culture, desire, and faith. To understand perceptual processes requires examination of all these facets of human behavior" (p. 3).

In the particular context of attempts at landscape evaluation, K. D. Fines (1968) has outlined: (i) the complexity of stimuli transmitted by landscapes to all, or a certain number of the five senses; (ii) the fact that individuals compare different landscapes and also appraise any given landscape in differing ways according to varying atmospheric conditions; and (iii) that their responses are conditioned by both inborn and culturally acquired characteristics. Faced with such a formidable array of problems it is not surprising that all methods for evaluating rural resources in general, and landscapes in particular, remain at an experimental stage and are thus open to considerable criticism. The following discussion will outline some of the approaches that have been pursued so far.

IDENTIFYING LANDSCAPE COMPONENTS

A variety of data is available to help identify land-use variations. S. B. K. Clark (1968) listed air photographs, topographic maps, and special sheets (showing land use, geology, and vegetation) as valuable sources to supply information on spatial variations in landscape characteristics. D. L. Linton's (1968) study of Scottish scenery as a natural resource exemplified a geographical approach to the problem. Six main types of "landform landscape" were designated, along with seven types of "land-use landscape" (or cultural landscape). Each tract of country was given a score on both scales. A composite assessment for sub-regions was then derived by combining the scores from both scales. This approach had the merit of being cheap and quick to undertake, since it made use of existing cartographic sources and did not involve investigation in the field. It differentiated landscape sub-regions for a large area (the whole of Scotland) but did not really tackle the problem of evaluating the varying qualities of the types of landscape identified.

A reconnaissance survey for the Nottinghamshire/Derbyshire sub-region sought to identify variations in landscape character by recording dominant features of ground profile, tree cover, field boundaries, and agriculture for almost a thousand regularly spaced sampling points which were investigated in the field. The planning group considered that analysis of these four characteristics would give an adequate representation of landscape differences. The results were recorded as an overlay for a 1:63,360 topographic map. Marked "edges" or "breaks" in rural landscapes were also noted. No attempt was made to evaluate the different types of landscape that were identified.

Landscape architects have taken a different approach to the task of identifying landscape components. A. E. Weddle's (1969) study of the land around the

Clyde estuary in Scotland illustrates their types of technique. Weddle was commissioned to make recommendations on the likely visual effects of industrial development at various locations in the study area. The basic problem was to choose the least intrusive site for future large-scale industry, including steelworks, oil refineries, and power stations. Weddle emphasized that a clear distinction could be drawn between "inherent landscape quality" (or the factual contents of each landscape) and the "acquired value" of individual landscapes. He insisted that "a landscape is of value only when it is of some use to the community" (p. 387). At this stage "inherent landscape quality" will be considered. "Acquired values" will be reviewed later. Four major pictorial components of landscapes were described: viewpoints and foreground; view; background; and lighting. These were divided into ten sub-categories, ranging from land use and landforms, to "framing", "edge effects", and prevailing climatic effects. Thus a far more complex approach to landscape description for a small area had been adopted than that used, by Linton for example, for a whole country. Clearly, the landscape architect's line of attack would only be feasible for limited stretches of terrain and would require the application of architectural or artistic skills in the necessary fieldwork.

The National Parks Commission's changing countryside pilot survey (1968) employed a much simpler version of a similar technique in an attempt to obtain samples of factual information on visually distinct components in rural landscapes for 1 km² grid squares throughout England and Wales, disregarding any considerations of genesis or function for such components. The survey paid particular attention to distinctions between water, various forms of vegetation cover, the presence of tall features, scattered trees, and edge effects in each grid square. Each of the operations considered so far has essentially been descriptive in character. As yet no attention has been paid to the actual evaluation of landscapes.

EVALUATION TECHNIQUES

Evaluation techniques involve devising methods to obtain individuals' reactions to landscapes by requesting them to identify and rank their preferences. A frequently used market-research technique is to display a range of photographs, here showing different landscapes, and then to obtain a ranked order of preference from each individual interviewed. Work undertaken by K. D. Fines (1968) and other members of the county planning staff in East Sussex illustrates the problems involved in such an approach. Fines selected a sample of forty-five people to rank photographs of twenty different landscapes. In fact he only used the results offered by the ten individuals in the sample who had considerable training in and experience of design. This decision was justified on two scores. First, such people were considered most likely to seek and to obtain the greatest enjoyment from landscapes; and, second, the wide scale of values associated with that group's responses, it was hoped, would represent the experience of the majority in the future as standards of education were improved and lengths of leisure time increased.

This type of decision to select only a special group of respondents raises the serious problem of whose opinions should be heeded, since individuals and sub-groups in society appraise landscapes in quite different ways. Conservation and landscape preservation have been shown to be essentially an upper middle-class social movement in North America, with very little appeal to working-class groups (Harry *et al.*, 1969). Some individuals prefer recreation environments which compensate for their normal living conditions. A proportion of urban dwellers, for example, would prefer "unspoilt", undeveloped landscapes, and would evaluate these very highly. By contrast, others would choose more "developed" landscapes with which they were familiar and in which they felt more at home.

The basic evaluation of landscape quality by individuals with design training was derived from photographs and was then transferred into the context of East Sussex by K. D. Fines. In other surveys, such as those in the New Forest (Hampshire) and Hertfordshire, the task of evaluation has been undertaken by members of the planning staff. Very rarely have the opinions of the man in the street actually been taken into account, even though it is ultimately for his future well-being that landscapes will be either conserved or developed. This situation stems partly from the enormous difficulties involved in attempting to interpret the preferences exhibited for specific photographs and then to translate these preferences into the particular context of the area being evaluated. Workers in the United States have undertaken detailed statistical analyses of reactions to photographs of rural landscapes in an attempt to identify particular landscape features and combinations of features that are repeatedly preferred by members of the public. Results of the investigation showed that quantitative expressions could be derived to represent significant combinations of features that were preferred when the photographs were ranked. These expressions included such conditions as "the area of intermediate vegetation multiplied by the area of distant non-vegetation" (Shafer *et al.*, 1969, p. 14). It is highly unlikely that members of the public are aware of such complications as they make their landscape preferences.

Another way of approaching the difficult task of evaluation is to assume that the extent to which people enjoy a rural environment is reflected by the volume of use they make of it. Weddle's definition of "acquired values" for the Clyde estuary study came into this category. His technique was to survey the number of people using landscape units in the study area and to take checks on their socio-economic characteristics and whether they were day visitors or were visiting the region on a longer-term basis. Such investigations permitted the computation of "acquired values". These were then summated with the results of the "inherent landscape quality" study to produce an "aggregate landscape value" for each small tract of countryside.

Practitioners of cost–benefit analysis would derive monetary values for landscape components. However, S. B. K. Clark has suggested that it is not really the total landscape or the pleasure derived from it that is being evaluated by cost–benefit analysis, but rather a summation of the agricultural, silvicultural, and other resources located in the particular stretch of country. F. Medhurst (1968) has pointed out that "It seems likely that intangibles, such as the values of the components of a good environment—rich in vegetation, distant views, agreeable climate, etc.—will be the last to yield to this type of analysis. This is because these factors impinge on the human body, on the mind and the spirit, and whilst it is generally appreciated that these characteristics of the environment are desirable, an evaluation of them probably requires research in fields not yet involved in planning" (p. 61).

No matter which approach is adopted towards defining rural resources and evaluating them, the difficult stage arrives when information from a series of "single resource surveys" has to be synthesized. C. Steinitz (1970) has employed sophisticated techniques of computer mapping to reduce material derived from a variety of types of survey in New England to a common cartographic form of representation for visual comparison. Unfortunately this does not overcome the underlying problem of the inequality of class intervals and value judgements involved in each of the "single resource surveys". On other occasions the results of single resource surveys have been shown on a series of transparent overlays which can be superimposed to provide a cartographic "sieving" device. In the case of the east Hampshire area of outstanding natural beauty study, detailed surveys were made, first, of the area's main resources (wildlife/ecology, agriculture, forestry, and landscape) and, second, of the nature and extent of the use of these resources: 136 small resource zones were defined within the AONB. An acceptability chart or matrix was drawn up to allow a rapid identification of areas where land-use conflict might arise if suitable management techniques were not devised and applied.

A rather similar survey method was used in the Hertfordshire Countryside Plan (Kitching, 1970). Surveys were undertaken in 1968 across 196 tracts of country, each with fairly uniform internal characteristics, which together covered the whole county. Components within the landscape, such as woods, open land, water, buildings, and "eyesores", were allotted scores by the planning team on the basis of the following three factors: their visual contribution to or detraction from the landscape; their accessibility and contribution to recreational resources; and their value in relation to the rural economy. Maps of the county, divided into tracts with common values derived from the scoring system, accompanied each of the three analyses. The final Countryside Plan identifies various grades of landscape and rural resource and then defines areas which might be managed as regional parks, country parks, or "special country areas" for picnicking, walking, riding, camping and sight-seeing.

In so far as "the ultimate purpose of landscape science is to prepare the theoretical basis of how a cultural landscape can be developed" (Isaczenko, 1970, p. 731), the various surveys of landscape and rural resources have met with success. Individual planning authorities have acquired information on spatial differences in local landscapes and have attributed "values" to these landscape types. The Clyde estuary survey, for example, determined the relative losses which would result from installing heavy industries in various parts of the region. The landscape resource survey in East Sussex helped the county planning staff to reach decisions on where permission for urban expansion should be granted and where it should be withheld. In addition, the specially devised landscape map helped them to select routes for supergrid electricity lines (with pylons of heights of up to 70 m) and new roads in such a way as to minimize the landscape erosion which would inevitably result. As K. D. Fines (1968) has noted, land-use planning

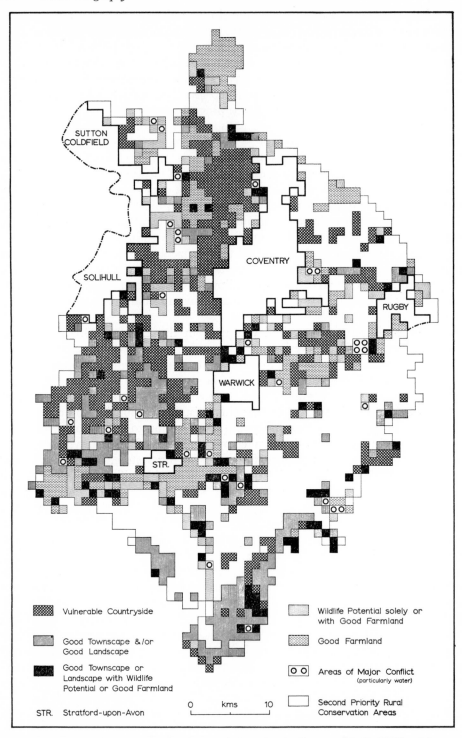

Fig. 9.1. Rural resource evaluation carried out for the Coventry/Solihull/Warwick-shire sub-regional planning study.

mechanisms have been developed to attempt to safeguard national parks, areas of outstanding natural beauty, and peri-urban green belts, but "they have generally been inefficient in preventing the slow erosion of the landscape and townscape of the less privileged but greater part of Britain . . . much of which is of a high *relative* value because it is the more populous and accessible" (p. 52). Rural-resource evaluation studies on a county or sub-regional basis such as that for Coventry/Solihull/Warwickshire (Fig. 9.1) will help to provide a body of background data from which management proposals may be prepared to resist and perhaps even reverse landscape erosion in the future.

However, the various evaluation studies that have been prepared so far are open to criticism on two main counts. First, their design and execution is based on a whole series of decisions regarding categories and qualities of land use and landscape that are derived from highly selective and objective criteria. D. M. Brancher's (1969) criticism of the East Sussex survey as "subjectivism dressed as science" (p. 91) is relevant both to the nature of the problem of evaluating landscapes and rural resources, and also to the state of the art at present. The second criticism stems from the character of proposals for rural management (country parks, conservation areas, etc.) which have been derived from the evidence of the surveys. As the Countryside Commission reviewers said of the Hertfordshire Countryside Plan: "admirable though such policies are, the plan which emerges seems to be more a response to existing conditions than an attempt to anticipate and provide for future developments" (Davidson and Tayler, 1970, p. 44). It is the unenviable but essential task of planning authorities to try to predict future requirements for recreation, residence, new employment, service provision, and other activities in the countryside and then remodel the rural environment in such ways as to ensure that these requirements may be satisfied.

REFERENCES AND FURTHER READING

Conceptual problems involved in evaluating landscapes are discussed by:

LOWENTHAL, D. (1967) Environmental perception and behavior, *Department of Geography Research Papers*, **109**, Chicago.

The general need for landscape classification for planning purposes is presented by:

HOOKWAY, R. J. S. and DAVIDSON, J. (1970) *Leisure: problems and prospects for the environment*, Countryside Commission, London.

A controversial attempt at landscape evaluation is presented by:

FINES, K. D. (1968) Landscape evaluation: a research project in East Sussex, *Regional Studies* **2**, 41–55.

See also the rejoinder by:

BRANCHER, D. M. (1969) Critique of K. D. Fines: landscape evaluation, *Regional Studies* **3**, 91–92.

A geographical approach to landscape study is described in:

LINTON, D. L. (1968) The assessment of scenery as a natural resource, *Scottish Geographical Magazine* **84**, 219–38.

The approach of a landscape architect is outlined in:

WEDDLE, A. E. (1969) Techniques in landscape planning, *Journal of the Town Planning Institute* **55**, 387–9.

Problems and techniques of landscape survey are considered in:

CLARK, S. B. K. (1968) Landscape survey and analysis on a national basis, *Planning Outlook* **4,** 15–29.

MEDHURST, F. (1968) A method of regional landscape analysis, *Planning Outlook* **4,** 61–69.

NATIONAL PARKS COMMISSION (1968) *Changing Countryside Survey* (Handbook), Second Pilot Survey, London.

SKINNER, D. N. (1968) Landscape survey with special reference to recreation and tourism in Scotland, *Planning Outlook* **4,** 37–43.

Examples of landscape and rural resource evaluation for practical planning purposes are presented in:

ANON. (1968) *Rural Planning Methods (East Hants AONB)*, Countryside Commission, London.

ANON. (1970) *Conservation of the New Forest: a draft report*, Winchester.

ANON. (1971) Coventry/Solihull/Warwickshire sub-regional study, *Journal of the Town Planning Institute* **57,** 481–4.

DAVIDSON, J. and TAYLER, C. (1970) Hertfordshire Countryside Plan 1970, *Recreation News Supplement* (Countryside Commission) **2,** 43–44.

JACKSON, K. (1971) Notts./Derbys.: a sub-regional landscape survey, *Journal of the Town Planning Institute* **57,** 203–4.

KITCHING, L. C. (1970) *Hertfordshire Countryside Plan*, Hertfordshire County Council, Hertford.

SIENKIEWICZ, J. (1971) A synoptic review of the Coventry/Solihull/Warwickshire sub-regional planning study, *Recreational News Supplement* (Countryside Commission) **5,** 24–28.

East European views of landscape science and the cartographic recording of visual features are summarized in:

GERASIMOV, I. P. *et al.* (1970) Current geographical problems in recreational planning, *Soviet Geography* **11,** 189–98; also *Ekistics* **184,** 220–2 (1971).

ISACZENKO, A. G. (1970) Theoretical and practical landscape science: its purpose, *Przeglad Geograficzny* **40,** 731–2.

VEDENIN, YU. A. and MIROSHNICHENKO, N. N. (1970) Evaluation of the natural environment for recreational purposes, *Soviet Geography* **11,** 198–207; also *Ekistics* **184,** 223–6 (1971).

The application of computer mapping techniques to synthesize resource surveys in New England is discussed by:

STEINITZ, C. (1970) Landscape resource analysis: the state of the art, *Landscape Architecture* **60,** 101–5.

Numerical analysis of public reactions to photographs of landscapes is presented in:

SHAFER, E. L. *et al.* (1969) Natural landscape preferences: a predictive model, *Journal of Leisure Research* **1,** 1–19.

Contrasting hypotheses of grouping or isolation; compensation or familiarity; are reviewed to try to explain recreation preferences:

BURCH, W. R. (1969) The social circles of leisure: competing explanations, *Journal of Leisure Research* **1,** 125–47.

PETERSON, G. L. and NEUMANN, E. S. (1969) Modeling and predicting human response to the visual recreation environment, *Journal of Leisure Research* **1,** 219–37.

Varying social reactions to landscape and other conservation issues are considered in:

HARRY, J. *et al.* (1969) Conservation: an upper-middle class social movement, *Journal of Leisure Research* **1,** 246–54.

CHAPTER 10

SETTLEMENT RATIONALIZATION IN RURAL AREAS

PATTERNS OF CHANGE

Rural settlement patterns in long-settled parts of the world have been inherited from the past when resident population totals were greater than they are today and the service needs of countryfolk could normally be satisfied by local crafts-men and traders operating in virtually every village. This situation has changed drastically in many rural areas over the last 100 years. It is true that population figures have been sustained, or may even have grown, in "metropolitan villages", but in more remote rural regions the number of country dwellers has declined and the demand for local services has contracted. Craftsmen have been unable to compete with factory producers, and many country shops have closed, leaving numerous small settlements with only a post-office-cum-general-store as their total commercial facilities. Some small villages are now without services of any kind.

Marion Clawson (1968) has outlined the nature of such problems in depressed rural areas: "A great many of the small rural towns* will wither greatly, and some will die and disappear by 2000. Some have already died and disappeared in farming areas. The process within and between such towns will be cumulative, in a downward spiral. As their business volume declines they will be less able to offer services that will attract farmers, who will gradually go to larger towns at farther distances. The decline of one kind of store or service will tend to depress the demand for the others; they are all linked together to a degree, since many trips to town are multi-purpose" (p. 287). The process in the countryside "will be one of continued decay, not of readjustment or of remodelling. As such it will not be pretty. . . . The process is never a happy one: many social ills arise out of it as the population ages, becomes less prosperous, and as all manner of social institutions decline. Perhaps above all, the spirit of such communities is nearly always depressed" (p. 288).

Some of the problems of rural isolation have been partly overcome by rising rates of personal mobility, especially through car ownership, as Clawson suggests. Most, but not all, country dwellers can drive to larger villages or to towns in order to purchase the goods and services that are not available in their home settlements.

* The term "settlement" might more readily be used in the general context of this discussion.

The costs of providing education, health, and other services for a declining and relatively dispersed population, in small villages, isolated hamlets, and farms are particularly high. In addition country dwellers now expect to receive supplies of electricity and piped water and to be connected to mains drainage systems. Again such utilities can only be provided at very great expense.

Clawson (1968) has eloquently summarized the situation in rural North America, where "the settlement pattern is literally 'horse and buggy', wholly unsuited to modern life" (p. 3). The same conclusion is equally valid for many rural areas in the Old World. However, the settlement patterns in such areas owe their origins to a medieval scale of economy and mobility rather than to one dating only from the nineteenth century. Rural planners now face the difficult task of devising programmes for rationalizing the provision of services and the structure of settlements in country areas where further declines in population, and hence in demand, may be anticipated in the future.

Numerous studies of population changes in country areas echo the words of D. Lowenthal and L. Comitas (1962) that "the smaller the . . . social unit, or community, the more likely it is to be losing population" (p. 196). A critical size threshold to separate settlements which are losing population from those which are retaining or even increasing it may not be recognized. Conditions vary greatly from region to region, but the general principle enunciated above appears to hold true. Detailed analyses of population changes in north-east England between 1951 and 1961 showed that settlements with more than 450 adult residents exhibited fairly consistent increases during that decade (House, 1965; Edwards, 1971). Smaller settlements generally experienced decline, with the loss of population and disappearance of social and commercial facilities being most severe in settlements with less than 100 adult residents. Similar conclusions were reached by R. J. Johnston (1965), working in north Yorkshire, who found that "the settlement pattern, particularly village size, has influenced the degree to which an area has retarded depopulation . . . and has been able to attract new, non-primary population" (p. 292).

G. Hodge's (1966) work on shopping and other commercial facilities available in small settlements in parts of North America takes the discussion a stage further by showing how settlement components in the trade-centre hierarchy change in importance in response to population decline in unplanned situations. His results showed the following points. First, trade centres with small populations were most susceptible to decline. Second, trade centres which offered a small range of goods and services were more likely to decline than those which offered a wide range. Third, small trade centres located with 16–24 km of larger towns tended to be less viable than if they were located in more remote areas beyond immediate urban hinterlands. Finally, two types of service centre were emerging— a large number of small centres which served only local needs, and a very small group of larger centres which met the specialized shopping requirements of residents in broader rural hinterlands.

Country planners in many parts of the world have attempted to pay due attention to these "automatic" trends as they formulate policies for future service provision in the countryside. Such policies in a given rural area involve selecting a limited number of settlements where population numbers may be increased and the range of service provision extended, and designating the remainder for future stability or even contraction. R. J. Green (1971) in his study of country planning in Britain maintains that "the primary objective of any rural plan is to secure changes in the pattern of settlement that will increase the range of social, commercial and public services, and education and employment opportunities, and make them available to a greater proportion of the resident population" (p. 85). Marion Clawson (1968) argues along a similar vein. "The abandonment of many small rural towns should be planned for and given assistance. . . . If the long-run future of the smallest towns is very dark, as it seems to be, a planned and aided withdrawal would be not only kinder to the people involved but would also minimize the social costs to the nation as a whole. At the same time, those larger towns that do have reasonable prospects for viability should be helped to grow" (p. 84).

G. Hodge's (1966) predictions of likely changes in trade centres in North America suggest the broad generalized trends of "automatic" modification in settlement structure that will take place throughout the developed world. First, the number of tiny service centres serving the farming population will decrease even further as farms are enlarged and greater mechanization takes place, thereby changing the man:land ratio and reducing the market potential of smaller centres. Second, many hamlets meeting daily shopping needs will disappear as most country people choose to drive to larger centres offering a wider range of specialized goods and services. Third, except for suburbanized settlements, small service centres around large towns will probably disappear within 16 km radii and will show substantial decline in areas up to 24 km away. Only beyond this distance is the trade-area integrity of small centres likely to remain secure. Fourth, as the settlement hierarchy is thinned out, rural people will have to travel much further to obtain day-to-day necessities. Higher-order settlements will tend to emerge in a more regularly spaced pattern to serve demands created by the enlarged incomes of farming households and their ability to exercise choice because of increased personal mobility.

The planned reorganization of rural settlement has the following wide-ranging objectives: (i) fuller utilization of rural capital assets (roads, public services, schools); (ii) concentration of traffic movement on to a limited number of primary road and rail routes; (iii) correlation of schemes for providing health services, schools, shops, roads, and public services; and (iv) the operation of a viable programme of employment growth or redistribution, based on industrial or commercial enterprise, where necessary stimulated by financial or other incentives (Green, 1971, p. 88).

Unfortunately little theoretical work is available to guide rural planners as they attempt to realize such objectives by determining how the process of "natural selection" between rural settlements might be speeded up or in which directions it might be steered so that scarce resources are not wasted on bolstering up services in small settlements that will continue to experience drastic population losses in the future. An interesting analogy to settlement changes has been drawn by J. C. Hudson (1969) who took the concept of "competition" from studies in plant ecology and applied it to the process whereby small service centres in rural areas are pruned away as increased personal mobility permits broader hinterlands to be served from a reduced number of "key" centres.

EXPERIENCE OF SETTLEMENT RATIONALIZATION

County Development Plans in England and Wales, prepared in accordance with the Town and Country Planning Act of 1947, have defined varying types of settlement, distinguishing those to be expanded as local service centres from others where only limited expansion would be permitted. Thus, for example, 100 centres out of 440 in the Lindsey division of Lincolnshire were designated as key villages.

This kind of approach had been adopted in the Cambridgeshire Village College system before World War II. This depended on the selection of a limited number of larger villages in which education and social facilities would be concentrated. This policy has continued in that county since the war with the building of further village colleges. Similar policies have been implemented in many other parts of Britain. John Saville (1957), working in the South Hams district of Devon, showed that most villages were far too small to be considered as nuclei for satisfactory social or economic living in the future. He argued along now familiar lines for a limited number of key centres. Smaller settlements would survive as commuter satellites around such key centres, but very remote villages would have to face decline in the future. Such a policy has been implemented, and the Analysis of the Survey of the Devonshire County Development Plan (1964) suggested that a thriving rural community should contain the following facilities (Table 16). To these one might add street lighting, recreation facilities, public transport, and safe roads with good surfaces.

TABLE 16. RECOMMENDED FACILITIES FOR A THRIVING RURAL COMMUNITY

Public utilities: mains water, electricity, sewerage
Social facilities: primary school, place of worship, village hall, and possibly doctor's surgery
Shops for day-to-day needs and a post office
Employment in the village or available conveniently nearby.

Source: The Analysis of the Survey for the Devonshire County Development Plan (1964).

More recently the Development Commissioners (1966) have spelled out the concept of trigger areas where investment and development would be concentrated. "The selected areas will be centred on growth points, which can be expected to prosper as attractive sites for additional industry, trade and population. The object of their initial favourable treatment is that it may start off a process of development to which further independent enterprises, not necessarily of an industrial character, will subsequently contribute" (p. 9).

Work on settlement planning in Norfolk (East Anglia) illustrates the problems involved in remodelling settlement hierarchies. The county contains 500 villages and hamlets, of which 80 per cent are losing population. The remainder are

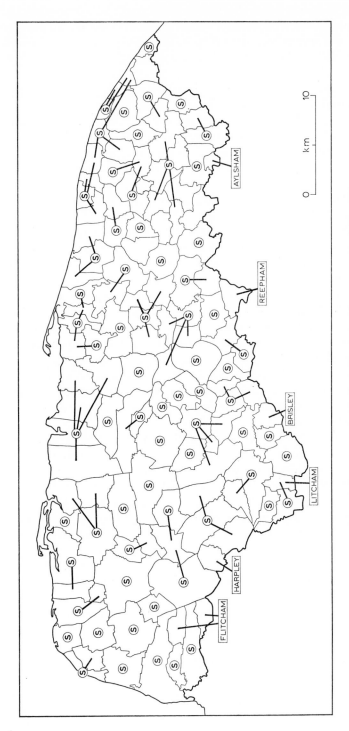

FIG. 10.1. Primary schools and their hinterlands in north Norfolk, 1970.

expanding or holding their own, often as a result of functioning as commuter settlements around employment centres such as Norwich or King's Lynn. But the commuters who settle in these villages are highly mobile and often look to their nearest large town rather than to their home village to meet their service requirements.

The settlement pattern of Norfolk involves a fairly regular spacing of hamlets and villages which are normally about 3 km apart. Three-quarters of these settlements contain fewer than 500 people and only 10 per cent have more than 1000. The agricultural labour force has decreased rapidly and will continue to contract in the future. Already the agricultural *raison d'être* of many settlements has virtually disappeared. Farming and its ancillary activities will be able to provide employment for only a small proportion of school-leavers in the future. Some village jobs are available for boys with secondary modern school training, but there are few jobs for girls and for grammar school trained boys.

Following the Education Act of 1944, basic schooling facilities have been withdrawn from some small villages, and children have to travel to neighbouring settlements. Green (1971) has noted that as the tendency to close village schools and concentrate all but infants' classes at a smaller number of centres continues, "daily travel out of the village is likely to become a regular habit from the age of 8 or 9" (p. 12). In the 1960's many village schools were housed in old buildings and served only small numbers of children. It was becoming increasingly difficult to attract ambitious teachers to such schools. Secondary schools, with more specialized teaching facilities, could only operate to serve more populous catchments, and were already located in the larger settlements of the county. Figures 10.1 and 10.2 for north Norfolk show the distinction in 1970 between primary school catchments, of perhaps one or two parishes, and the much broader catchments of secondary schools.

Most rural parishes in the county still contained a general store which, one suspects, was often run more as a way of life than as a profit-maximizing operation. Figure 10.3 shows that there were marked contrasts in the range of shopping facilities available between the average village with only one or two stores and the small market towns which possessed a number of types of store. Some parishes had even experienced the closure of the village pub (Fig. 10.4). Such a trend of service decline is typical of many rural areas which are beyond the commuting hinterlands of urban centres.

Planning staff in Norfolk and elsewhere have used the concept of the "range of a good" as they try to determine threshold levels for the operation of specific services. These will help them choose the number and spacing of centres for expansion and the target population that might be aimed for in rural hinterlands around key centres. The threshold for each good is defined as the minimum population needed to support that good, and the range of a good as the maximum distance a population is prepared to travel to a centre to obtain that good. As has already been mentioned, empirical studies show that larger settlements normally contain a wider range of services than smaller ones. But, unfortunately there is no simple correlation between a centre's population and either the size of the rural population it serves or the dimensions of its hinterland. For example,

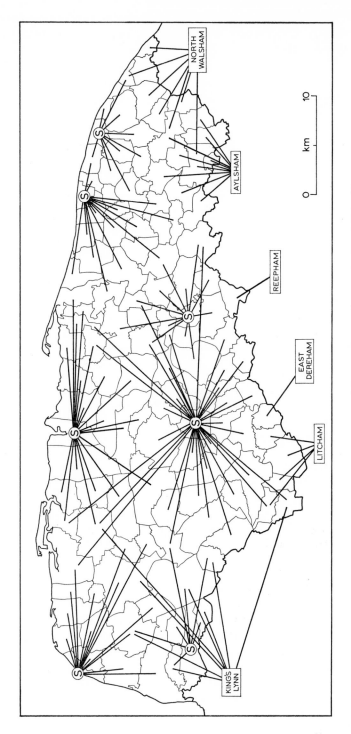

Fig. 10.2. Secondary schools and their hinterlands in north Norfolk, 1970.

industrial or mining settlements set in the midst of rural areas are often under-equipped by comparison with nearby, longer-settled market towns of equal population size (Edwards, 1971, p. 267).

TABLE 17. FACILITIES IN CAMBRIDGESHIRE VILLAGES

Number of inhabitants	
170–600	Public house, post office, hall, and general store
600–1100	Also a primary school, playing field, and garage
1100–1800	Also a police house, butcher, ladies' hairdresser, and doctor
1800–3000	Also an electrical goods shop, licensed club, hardware store, and barber
3000 plus	Also a secondary school and chemist

Source: Cambridgeshire and the Isle of Ely County Council Development Plan Revision, 1968; Report of Survey (1968).

Nevertheless, Table 17, derived from the existing services provided in rural Cambridgeshire, suggests the kind of relationship frequently found in the countryside with the range of services provided increasing as the resident population of settlements increases. Table 18 gives the critical values for service provision in rural Norfolk obtained during the 1960's. A primary school with a two-form

TABLE 18. MEAN NUMBER OF RURAL INHABITANTS PER
TYPE OF SHOP IN NORFOLK

	Inhabitants		Inhabitants
Grocer	300	Draper	2500
Butcher	2000	Household goods	2500
Baker	3000	Chemist	4000

Source: GREEN, R. J. and AYTON, J. B. (1967), Changes in the pattern of rural settlement. Mimeographed paper for the Town Planning Institute Research Conference, p. 4.

entry with classes for each yearly age group required a total hinterland population of about 5000 people from which to draw pupils. A secondary modern or comprehensive school required double that amount. A doctor working by himself normally served about 2000 patients, and hence a three-doctor group practice would require 6000 plus. Chemist's shops in rural parts of lowland England usually served at least 4000 potential clients. In short, R. J. Green and J. B. Ayton (1967) considered that "a minimum population of about 5000 is needed to support a reasonable range of local facilities and this suggests a need for a policy aimed at the expansion of a few selected villages and the acceleration of decline in the remainder" (p. 1).

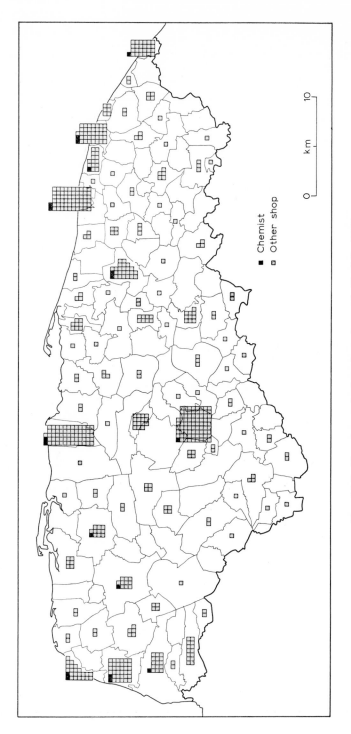

Fig. 10.3. Chemists and other shops in north Norfolk, 1970.

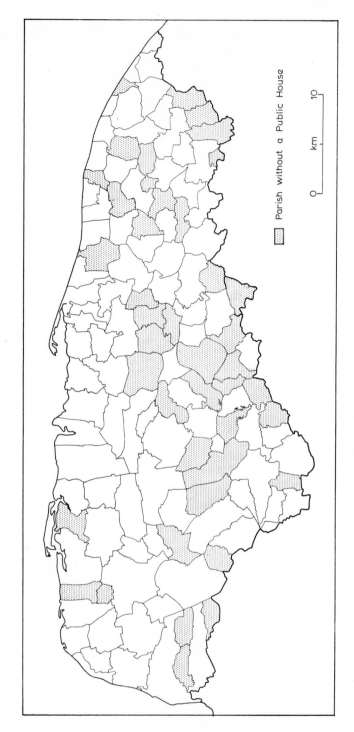

Fig. 10.4. Parishes without public houses in north Norfolk, 1970.

Such methods of analysis merit some comment. First, they apply only to the specific geographical milieu of eastern England with its particular arrangement of inhabitants in towns, villages, and hamlets distributed across lowland terrain. Second, they have been derived from considerations valid at a particular point in time, in this case the 1960's. One cannot assume, automatically, that such thresholds will still be meaningful in, say, 20 years time. It is interesting to note that 20 years ago villages with more than 500–600 inhabitants were considered as constituting viable rural settlements in south-west England (Bonham-Carter, 1951; Mitchell, 1951). Third, thresholds are valid only in their own specified frameworks. Studies in other countries with different power for planning intervention and administration (e.g. in eastern Europe) may provide very different critical values. Fourth, the thresholds relate to a given set of institutions (English primary schools, three-doctor practices, etc.). It is clear that threshold values would vary considerably if the institutional criteria used were changed. In addition, one must note that threshold values will continue to change through time (as they have done in the past) as rural residents decline in numbers, acquire more cars or deep freezes, or change their characteristics in any other way.

This final point may be illustrated in the context of the Dutch polderlands. Planned settlements, established in the north-east polder in the 1940's involved village regions of 4000 ha, which covered much larger areas than the average village region of 1200 ha on the mainland. Each new "region" was planned to contain about 3000 inhabitants of whom half would live in villages and the remainder on dispersed farms. This scale of operation did not prove successful. Many village regions failed to attract 3000 inhabitants. Farmworkers were willing to travel greater distances to work than was previously believed, in spite of intensive sociological investigation. Plans for the later and larger East Flevoland polder proposed twelve rural centres in 1954 (Fig. 10.5), but since then ideas have had to be changed drastically in the light of the experience of the north-east polder. M. Van Hulten (1969) has shown how the number of centres has been reduced to only four with population targets of 5000 living in village regions of 9000 ha. These are composed of much larger farms than on the earlier polders. In keeping with the increasing personal mobility, the maximum distance between farm and village has been increased from 5 km on the north-east polder to 12 km on East Flevoland. Even these adjustments have not proved completely successful since the regional centre of Lelystad has already usurped many functions from the middle-level centre of Dronten. Nevertheless, Thijsse (1963, 1968) could report that Dutch Government planners were appreciating the magnitude of recent changes in mobility and as a result were managing the new East Flevoland polder to meet likely future needs for recreation as well as residence.

County development plans in England and Wales have selected rural settlements for holding or even expanding population, but relatively little has been done to "accelerate the decline" of small settlements. Clearly this is a difficult operation politically since the inhabitants of small settlements do not wish to have their services actively phased out even though they may have to accept losing them through "natural selection" between settlements. R. J. Green and J. B. Ayton (1967) have developed this point in the East Anglian context: "Our studies indi-

R.G.—F

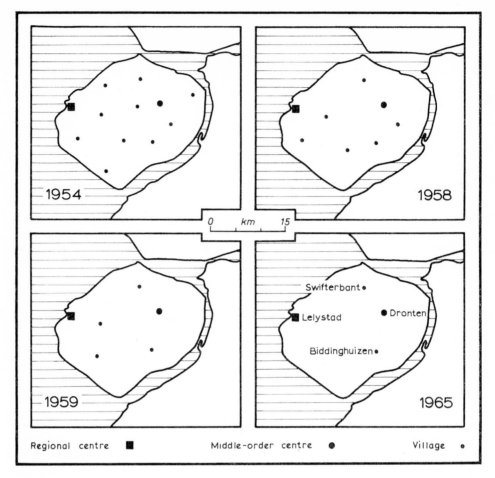

FIG. 10.5. Changes in plans for settlement development in the East Flevoland
polder, 1954–65.

cate a considerable mobility of population in rural areas, which seems to offer
scope for change, while the existing capital commitment in public services, schools
and housing is less likely to inhibit change than the 'social' commitment, real or
imaginary, inherent in existing small communities" (p. 1).

It may be relatively easy for planners to decide to build up services in central
settlements, but to deny growth to small settlements is a difficult decision to take
and is even more difficult to implement. In the English context, this would mean
constructing no more council houses in villages designated for contraction and
refusing permission for speculative house building. Figure 10.6 suggests what this
might mean in spatial terms. Services, such as the village school, pub, and shop,
would be withdrawn. Some compensation might be provided by travelling shops
during a transition phase, but eventually they would be withdrawn. "Probably
the hardest problem to solve will be the provision and maintenance of facilities

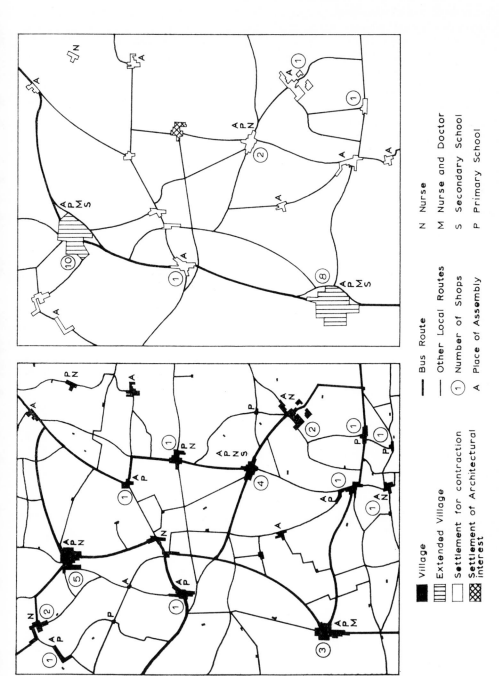

FIG. 10.6. Hypothetical changes in rural settlement in eastern England as a result of rationalization.

and services in the declining villages, until these become, in the future, what are accepted as hamlets today" (Green and Ayton, 1967, p. 7).

Not all schemes for building up "key centres" have also involved suggestions for the progressive abandonment of minor settlements, although this has sometimes been feared. In the Limerick Rural Survey (1964) J. Newman reported on popular reactions to his earlier proposals for strengthening chosen country towns for the reinvigoration of rural Ireland. "This recommendation was misinterpreted by some, who feared that it meant the abandonment of the small villages in favour of the towns. The writer wishes to emphasize that this is not so. . . . It would be necessary for groups of such villages to 'club together', so to speak, in wider units for purposes of employment and social provision, by linking themselves in what might be called 'satellite capacity' around selected strong centres, that is, rural towns" (p. 248).

By contrast, plans for settlement modification in County Durham have involved the liquidation of some non-viable villages. Admittedly these were small mining settlements, rather than agricultural villages, but the basic issue to be faced by planners was similar. The large number and small size of many mining villages in the Durham coalfield was recognized as a planning problem immediately after World War II. The first county development plan designated four types of settlement (Table 19; Fig. 10.7). Since then a concerted effort has been made to encourage the concentration of new housing development and investment in a limited number of places. In 1964 the revision of the county plan provided a rather more flexible approach and defined six types of settlement instead of four. Category D villages were subdivided into those where no development would be allowed, but settlement would continue to exist for a long time; and others where no new development would be allowed to occur and the clearance of houses would take place. One-third of the 370 settlements in the county are considered to have no long-term future, but collectively they contain less than one-tenth of the total population. The elimination of small villages from the settlement pattern has been a gradual process, so that by 1970 only eight small settlements had been completely cleared and their sites assimilated into the surrounding countryside.

Planners in other parts of Britain have propounded similar proposals for pruning settlement hierarchies and concentrating services at fewer points in the countryside. The Highlands and Islands Development Board, for example, has

TABLE 19. SETTLEMENT CATEGORIES IN COUNTY DURHAM

Category	
A	Population expected to expand
B	Population expected to remain stable
C	Population expected to be held at a reduced level
D	Population expected to go on shrinking

Source: SENIOR, D. (n.d.), *Growth Points for Durham*, Durham County Council, Durham, p. 10.

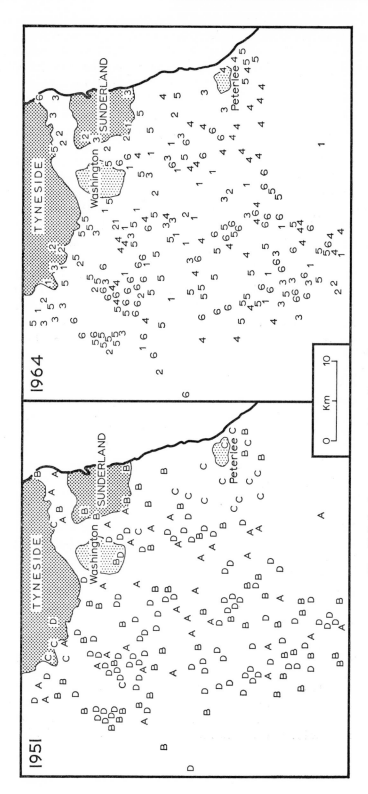

Fig. 10.7. Settlement categories in County Durham, 1951 and 1964.

favoured the idea of improving services and employment facilities at twenty-five centres in north-west Scotland. But, given the special problems of spatial isolation in this region, a wide variety of hinterland sizes has had to be accepted. These deviate enormously from optimal values, such as those quoted for Norfolk. Some of the Highland hinterlands contain populations ranging from 6000 to 16,000, but the majority have only between 500 and 3000. R. J. Green (1971) reports that a group of British planners when faced with the task of selecting an optimum size for viable rural settlements in the 1960's quoted a wide range of population sizes from 500 up to 8000.

In short, there is little agreement on the ideal size for "key villages". The values for one region, derived from considerations of settlement distribution, local terrain, ease of transportation, and administration, cannot be transferred automatically to another region, e.g. from the lowlands of East Anglia to the uplands of Wales or Scotland, or from the context of Britain to that of the USSR. Some researchers, studying problems of settlement rationalization in upland areas with dissected terrain, have questioned the meaningfulness of the whole threshold concept in such areas of low population density and difficult communication. Thus J. G. Thomas (1972), working in central Wales, found that "economies of size become impossible when the size involves population numbers which could be adequately serviced by one police officer, or doctor, or teacher, but which occupy such a large terrain that such an arrangement becomes physically impossible" (p. 101).

A CASE FOR PRAGMATISM?

One might argue that all rural areas that are losing their resident population face similar problems, but it would be erroneous to consider that such problems are identical and can be minimized by the application of identical planning tactics. East European literature on settlement modification exemplifies this point. S. A. Kovalev (1968) has pointed out that "the programme of the Soviet communist party [1961] lists the gradual elimination of social-economic and cultural-service differences between town and countryside as one of the major future tasks of the construction of communism" (p. 641). Similar points of view for reducing the rural/urban differential have been put forward in legislation in many other European countries. In the USSR this is to be achieved by (i) expanding large rural centres (where services, industries and cultural activities would be introduced) and running down smaller ones; (ii) introducing small-scale industry into rural settlements; and (iii) improving the planning and design of rural communities to provide them with amenities on a par with those in urban areas. No threshold figures are quoted in the general discussion, but elsewhere Kovalev and Ryazanov (1968) note that "Soviet policy is to encourage consolidation to optimal sizes of 1000 or 2000 people, to make possible better services than can be extended to single dwellings and to other very small settlements" (p. 651).

Soviet authors, and their western counterparts, plead first for the formal recognition of an hierarchy of central places; second, the allocation of each settlement to its appropriate level in the hierarchy; and, third, the equipment of each settlement with suitable facilities for the level to which it has been assigned. This final point would, of course, involve running down the very smallest units. A sixfold hierarchy (Table 20) has been proposed in the USSR, from the highest order of local service centre (the *raion* centre), through the centres of state and collective farms, to a variety of much less significant settlements. As has already been mentioned, central places serving between 1000 and 2000 inhabitants are frequently quoted as the ideal size.

TABLE 20. SETTLEMENT HIERARCHY IN RURAL REGIONS OF THE USSR

1. The *raion* centre
2. Central settlements of state and collective farms
3. Settlements of outlying farm divisions and work brigades
4. Local centres in areas remote from *raion* centres (e.g. livestock-feeding stations shared by collectives)
5. Specialized livestock farm settlements
6. Seasonal settlements

Source: KOVALEV, S. A. and RYAZANOV, V. S. (1968), Paths of evolution of rural settlement, *Soviet Geography* **9**, 660.

Thus in Moscow *oblast*, for example, the minimum population for maximum economy of servicing and social welfare should not be less than 1000 people. Only 850 of the 7500 rural settlements in the *oblast* will be expanded as "holding points" for the future. A further 2500 will be supported during a transitional phase up to 1980, but the remaining 4150 will be run down in the immediate future. Likewise, in the Belorussian SSR only 5500 of the 34,000 existing settlements will experience a progressive improvement in service provision in the future. Other examples of Soviet settlement replanning emphasize that, whilst the general policy for concentrating services at a smaller number of key centres can be applied in all rural areas, very different thresholds must be employed for areas with differing relief, population density, and economic activities.

Similar policies have been carried through in other parts of eastern Europe, being linked to the process of agricultural collectivization and administrative reform. In south-west Hungary, for example, the administrative pattern of Transdanubia has been reorganized into eighty-two micro-regions with mean populations of about 3000 inhabitants. Small settlements with under 300 inhabitants have lost their administrative autonomy and are having their services run down. Villages with over 1500 inhabitants are being strengthened and those over 5000 are being further developed towards "urban" conditions. Such a policy for liquidating minor settlements could not be implemented so readily under the more liberal regimes of the West.

Common threshold values do not emerge from a comparison of proposals for settlement re-planning in various parts of the world. This is not surprising since different values would be expected from the combination of the following factors in each specific rural region.

First, the degree of rural depopulation varies greatly from place to place and between different types of agricultural region. P. Scott's (1968) work on Tasmania has shown that trade centres at a given level in the settlement hierarchy perform rather different functions depending on whether they are in wheatland or cattle country or in areas of intensive or extensive farming.

Second, the social character of rural areas varies enormously. Some are still largely agricultural, others house retired urbanites, part-time farmers, weekenders, or holidaymakers. Different demands on rural services will thus be made.

Third, individual communities are very different in their sociological characteristics, as G. D. Mitchell (1951) has shown.

Fourth, the degree of settlement nucleation and dispersion varies enormously between localities.

Fifth, terrain conditions also vary. Inhabited areas in regions with deep valleys and high ridges or with islands and desolate coasts will be far more isolated than in lowland areas with much easier transportation.

Sixth, local communication facilities, in the forms of public and private passenger transport, mobile shops and other services, vary between regions, and from country to country.

Seventh, financial assistance for removing rural/urban inequalities, for example, by subsidizing services in the countryside varies from state to state.

Finally, differences in political organization appear to make it easier to rationalize rural settlement structures and the provision of services in eastern Europe and the USSR than in the West.

There are undoubtedly important local variations in the problems that have to be faced in the countryside; nevertheless, there are certain common requirements from country planning. Settlement structures and the provision of services need to be rationalized. Adequate forms of local employment and public transportation need to be guaranteed. In short, rural areas require management in an integrated, developmental fashion to cater for likely future needs.

REFERENCES AND FURTHER READING

Variations in the ability of different-sized settlements to retain population are discussed in:

BRACEY, H. E. (1952) *Social Provision in Rural Wiltshire*, Methuen, London.
BRACEY, H. E. (1970) *People and the Countryside*, Routledge & Kegan Paul, London.
BUTLER, J. E. and FUGUITT, G. V. (1970) Small-town population change and distance from larger towns, *Rural Sociology* **35**, 396–409.
CLAWSON, M. (1966) Factors and forces affecting the optimum future rural settlement pattern of the United States, *Economic Geography* **42**, 283–93.
CLAWSON, M. (1968) *Policy Directions for US Agriculture*, Resources for the Future Inc., Johns Hopkins University, Baltimore.
EDWARDS, J. A. (1971) The viability of lower size-order settlements in rural areas: the case of NE England, *Sociologia Ruralis* **11**, 247–75.
FUGUITT, G. V. and DEELEY, N. (1966) Retail service patterns and small-town population change: a replication of Hassinger's study, *Rural Sociology* **31**, 53–63.
HARDEN, W. (1960) Social and economic effects of community size, *Rural Sociology* **25**, 204–11.
HASSINGER, E. (1957a) The relationship of trade-centre population change to distance from larger centres in an agricultural area, *Rural Sociology* **22**, 131–6.

HASSINGER, E. (1957b) The relationship of retail-service patterns to trade center population change, *Rural Sociology* **22,** 235–40.

HASSINGER, E. (1961) Social relations between centralized and local social systems, *Rural Sociology* **26,** 354–63.

HODGE, G. (1965) The prediction of trade-center viability in the Great Plains, *Papers and Proceedings of the Regional Science Association* **15,** 87–115.

HODGE, G. (1966) Do villages grow? Some perspectives and predictions, *Rural Sociology* **31,** 183–96.

HOUSE, J. W. (1965) *Rural NE England: 1951–61,* Department of Geography, University of Newcastle-upon-Tyne, Papers on Migration and Mobility, **1.**

JOHNSTON, R. J. (1965) Components of rural population change, *Town Planning Review* **36,** 279–93.

JOHNSTON, R. J. (1966) Central places and the settlement pattern, *Annals Association of American Geographers* **56,** 541–9.

JOHNSTON, R. J. (1967) A reconnaissance study of population change in Nidderdale, 1951–61, *Transactions Institute of British Geographers* **41,** 113–23.

LOWENTHAL, D. and COMITAS, L. (1962) Emigration and depopulation: some neglected aspects of population geography, *Geographical Review* **52,** 195–210.

RIKKINEN, K. (1968) Change in village and rural population with distance from Duluth, *Economic Geography* **44,** 312–25.

SCOTT, P. (1968) Trade-center population change, centralization and trade-area farming type, *Rural Sociology* **33,** 424–36.

A regional example of retail organization in the countryside is found in:

TARRANT, J. R. (1967) *Retail Distribution in Eastern Yorkshire,* University of Hull Occasional Papers in Geography, **8.**

Changes in rural settlement size and function are compared with botanical processes in:

HUDSON, J. C. (1969) A location theory for rural settlements, *Annals Association of American Geographers* **59,** 365–81.

Thresholds and the minimum range of a good are discussed by:

BERRY, B. J. L. (1967) *Geography of Market Centers and Retail Distribution,* Prentice-Hall, New York.

GARNER, B. J. (1967) Models of urban location and settlement location, in CHORLEY, R. J. and HAGGETT, P. (eds.), *Models in Geography,* Methuen, London, pp. 303–60.

The village college system is discussed in:

WEDDLE, A. E. (1964–5) Rural land resources, *Town Planning Review* **35,** 267–84.

Proposals for settlement rationalization with particular reference to eastern England are analysed in:

DRUDY, P. J. and WALLACE, D. B. (1971) Towards a development programme for remote rural areas: a case study in north Norfolk, *Regional Studies* **5,** 281–8.

GREEN, R. J. (1966) The remote countryside: a plan for contraction, *Planning Outlook* **1,** 17–37.

GREEN, R. J. (1971) *Country Planning,* University of Manchester Press, Manchester.

GREEN, R. J. and AYTON, J. B. (1967) Changes in the pattern of rural settlement. Mimeographed paper for the Town Planning Institute Research Conference.

Changing ideas on desirable patterns and minimum support populations for rural settlements may be traced in:

BONHAM CARTER, V. (1951) *The English Village,* Penguin, Harmondsworth.

DAVIES, M. L. (1968) The rural community in central Wales: a study in social geography, in BOWEN, E. G. *et al.* (eds.), *Geography at Aberystwyth,* University of Wales Press, Cardiff, pp. 205–18.

HOFSTEE, E. W. (1970) The relations between sociology and policy, *Sociologia Ruralis* **10,** 331–45.

KOVALEV, S. A. (1972) Trnasformation of rural settlements in the Soviet Union, *Geoforum*, 9, 33–45.

MITCHELL, G. D. (1951) The relevance of group dynamics to rural planning problems, *Sociological Review* **43**, 1–16.

ORWIN, C. S. (1944) *Country Planning*, Oxford University Press, London.

THIJSSE, J. P. (1963) A rural pattern for the future in the Netherlands, *Regional Science Association: Papers and Proceedings*, **10,** 133–41.

THIJSSE, J. P. (1968) Second thoughts about a rural pattern for the future in the Netherlands, *Regional Science Association: Papers and Proceedings* **20,** 69–75.

VAN HULTEN, M. H. M. (1969) Plan and reality in the Ijsselmeerpolders, *Tijdschrift voor Economische en Sociale Geografie* **60,** 67–77.

The category D villages of County Durham are discussed by:

ATKINSON, J. R. (1970) Aspects of planning, in DEWDNEY, J. C. (ed.), *Durham County and City with Teesside*, Durham, pp. 433–42.

SENIOR, D. (n.d.) *Growth Points for Durham*, Durham County Council, Durham.

THORPE. D. (1970) Modern settlement, in DEWDNEY, J. C. (ed.), *op. cit.*, pp. 392–416.

Details of village management are contained in:

THORBURN, A. (1971) *Planning Villages*, Estates Gazette, London.

Examples of plans for rural settlement rationalization in the USSR and eastern Europe are found in:

KOLTA, J. (1969) Les préparatifs, la réalisation et les résultats obtenues de l'organisation des régions communales dans le Comitat Baranya (Hungary), in ENYEDI, G. and PALYANSZKY, P. (eds.), *Géographie et l'Aménagement du Territoire*, Budapest, pp. 169–95.

KOVALEV, S. A. (1968) Problems in the Soviet geography of rural settlement, *Soviet Geography* **9,** 641–51.

KOVALEV, S. A. and RYAZANOV, V. S. (1968) Paths of evolution of rural settlement, *Soviet Geography* **9,** 651–64.

Brief summaries of settlement rationalization in parts of the USSR are contained in *Geo Abstracts*, Series D, Social Geography, 1971; Belorussia (abstract 0183), Uzbekistan (0184), Moscow *oblast* (0199), and Transcarpathia (0200).

A review of the situation in the Republic of Ireland plus experience in France and the Netherlands is found in:

NEWMAN, J. (1964) *The Limerick Rural Survey*, Muintir na Tiré rural publications, Tipperary.

A critical consideration of the practical limitations of thresholds for service provision is contained in:

THOMAS, J. G. (1972) Population change and the provision of services, in ASHTON, J. and HARWOOD LONG, W. (eds.), *The Remoter Rural Areas of Britain*, Oliver & Boyd, Edinburgh, pp. 91–106.

The idea of the trigger area is outlined in:

DEVELOPMENT COMMISSIONERS (1966) *Aspects of Rural Development: 32nd Report of the Development Commissioners for the three years ended 31.3.1965*, House of Commons Session 1966–7, 100.

CHAPTER 11

MANUFACTURING IN THE COUNTRYSIDE

ADVANTAGES AND DISADVANTAGES OF INDUSTRIAL INSTALLATION

Many of the traditional craft industries and services which flourished in the countryside in the past have now disappeared. Nevertheless, efforts have been made over recent years to diversify employment opportunities in many predominantly rural regions by strengthening or introducing manufacturing and tertiary activities to sustain the remaining rural population. Such schemes for industrial development are sometimes linked to policies for key villages, holding points, or trigger areas.

Attitudes regarding the desirability of developing industries in rural areas are divided. The report of the Scott Committee (1942) stressed that both advantages and disadvantages would result from the installation of industries at selected points in the countryside. Five main advantages were identified. First, country life would be revived and town and country brought closer together as small towns were selected as holding points and key personnel were moved from urban areas to live in the countryside and manage or work in the new factories. Second, manufacturing industries would provide much-needed sources of employment for the wives and daughters of rural workers. Third, they would provide alternative avenues of employment for school-leavers, both boys and girls, who would normally have to look beyond their home areas in search of jobs. Fourth, industries might provide seasonal jobs for agricultural workers. Fifth, factory installation should encourage improvements in the physical amenities and social standards of the countryside. For example, schemes might be started for electricity, gas and piped-water supplies to be made available more widely. Similarly, facilities for education and recreation might be improved at chosen holding points. Industrial development would bring in extra rates for the local council, might bring more custom to shops, and also revitalize social activities. Factory installation in rural areas is frequently welcomed first by professional men and shopkeepers who anticipate an enlargement of their clientele, and, second, by working-class parents who are keen to ensure the provision of local jobs for their children.

Disadvantages might also result from industrial development in the countryside.

The Scott report noted that the following problems might threaten if new industries were not planned carefully. Productive farmland might be lost, agricultural activities dislocated, and farm units broken up by new industrial buildings or access roads. Harmful effects on agricultural production might stem from noxious fumes and effluents. Too many young people might be attracted away from farming to alternative forms of employment so that sufficient numbers of farm-hands might not be found. The beauty of the countryside might be spoiled by the bad siting and poor design of factory buildings. Finally, the social changes which would result from the impact of urban ideas and standards on rural mentalities might be considered undesirable. Undoubtedly social changes would take place, but their merits or otherwise would be a matter of debate. For example, a vocal managerial class would be installed with high demands for stimulating local authorities to provide improved housing, education, sewerage, hard roads, street lighting, and social facilities that would be to the benefit of all country dwellers involved. But, of course, such advantages might be seen as outweighed by the inevitable modification of traditional village life that would result.

Hence some groups in the countryside tend to be less than enthusiastic about industrial development. Farmers emphasize the possibilities of losing land and labour. Middle-class newcomers object to the likelihood of the rural way of life being destroyed, that they had left the city to find, or perhaps more correctly to mould into their own image. Some industrialists may themselves be hostile to rural industrial development, emphasizing the advantages of urban locations. Rather than continue the discussion at this level of generalization, an examination of a regional case study will pinpoint some of the practical problems involved in industrial development in the countryside.

INDUSTRIAL DEVELOPMENT IN CENTRAL WALES

The central part of Wales covers 40 per cent of the Principality but contains only 7 per cent of its total population. The number of local inhabitants fell from a peak of 275,000 in 1871 to 205,000 in 1971. In 1957 the five mid Wales county councils formed the Mid-Wales Industrial Development Association to focus their attack on the social and economic problems of the region which had manifested themselves through protracted rural depopulation. The Association is financed both by the counties and the Development Commission. Its role is to stimulate action by local authorities and the Government alike through bringing together industrialists, who wish to expand their productive capacity in the countryside, and localities which wish to develop factory installations.

The Association attracted fifty-three new industrial projects to the region between 1957 and 1969 (Table 21). Five more were attracted by 1972. Over forty were operational, employing 3500 people, and fifteen were under construction which will provide a further 1000 jobs. Thus the number of industrial jobs available in central Wales more than doubled in just 15 years. Thirty-three of the new factories were financed by the Government (Development Commissioners and Board of Trade) in an attempt to show a good example that private developers might follow. This was slow to start with but progress has been stimulated since 1966 by the scheduling of central Wales as a "development area" to receive special government finances for industrial installation.

The Association attempted to provide a reasonable distribution of new industrial development in the area's 21 small towns and spread the opportunity of industrial employment as widely as possible. Nevertheless, the secretary of the Association explained: "Even if desirable, to carry industry into every corner of the countryside is unlikely to be practical. The average industrialist looks for a location which can provide him with the services, schools, transport and amenities he requires: he certainly looks for an area where he is likely to attract the labour he needs, and he expects to see some sign of 'life' in the town if he and his family or those of his key workers are to live there. Industrialists are not much interested in places which are static" (Garbett-Edwards, 1972a, p. 53).

TABLE 21. NEW FACTORIES OPENED
IN MID-WALES, 1957–69

Clothing	9
Engineering	15
Packaging	2
Plastics	2
Office equipment	2
Others*	18
Advance factories to be let	5
Total	53

*From perfume to floor tiles.

Source: GARBETT-EDWARDS, D. P. (1972a), The establishment of new industries, in ASHTON, J. and HARWOOD LONG, W.(eds.), *The Remoter Rural Areas of Britain*, Oliver & Boyd, Edinburg, p. 55.

New industries are needed in the countryside to absorb a wide variety of types of available labour, ranging from redundant farm labourers and quarrymen to unemployed women and school-leavers. Rural areas are often keen to try to retain their bright young people and therefore seek to introduce enterprises involving skilled employment and research. This kind of development is more likely to be attracted when an industrial base already exists that can be built upon. Firms producing clothing, light engineering products, and other such goods are invaluable in creating such a base.

Indeed, only light manufacturing can normally be expected to establish in outlying rural areas. Such firms would probably be in the following sectors: engineering, packaging, electronics, plastics, office equipment, and food processing. It is advantageous to the reception area if incoming firms have a high demand for labour and offer training facilities for their new employees. Factories producing low-weight, high-quality goods are always important, since transport costs involved are low. However, it is now realized that office employment as well as factory work is needed in the region to reduce the numbers of bright young people leaving central Wales. The Association notes that property developers are not much interested and suggests that the Government should fill the gap by building advance offices in the same way factories are built ready for firms to occupy.

Three types of industrial plant have been installed in mid Wales: branch factories of well-established firms (twenty-seven factories); the total removal of firms to the area (sixteen); and the location of completely new companies (five). Branch factories offer the advantage that larger firms are involved which have greater resources to enable them to overcome the problems of development in the countryside. But, at the same time, branch factories have disadvantages, especially of threatening to be the first to close in times of economic difficulty.

It is frequently argued that local forest and farm resources should be processed in rural-based industries. The secretary of the mid Wales association noted: "It is usually assumed that such industry will reflect its benefits throughout the area's agriculture—and so it may in a horticultural or livestock-rearing area. In mid-Wales we have found such industry hard to identify and experience suggests that the output of areas of marginal farming (coupled with problems of distance from markets) is unlikely to be sufficient to attract processing firms" (Garbett-Edwards, 1972a, p. 56). Most of the manufacturing industries in the region in fact display little dependence on the material resource base of the area. P. Mounfield and H. D. Watts (1968) explain that "under favourable circumstances, the agricultural economy of a prosperous rural area can provide a source of raw materials, a market for other industries and various agricultural service trades may act as pivots around which manufacturing industry can grow. Much of upland agriculture in mid-Wales is not prosperous and the farms are heavily livestock-orientated. The creameries are the only representatives of the food industry" (p. 180). Tanning has survived at a few sites, as have a few timber firms. But new industries introduced in this region are not normally based on the processing of agricultural products.

Any effort to attract industry to a rural area must ensure that attractive industrial sites are available at the small towns or other settlements that have been chosen as suitable locations for industrial installation. Experience in central Wales suggests that the availability of existing buildings, such as church halls or old schools, is more important than simply providing a site. "It is the experience of most people concerned with the practical aspects of industrial development that every industrialist wants his new factory 'yesterday'" (Garbett-Edwards, 1972a, p. 58). Existing buildings may be leased. This enables industrialists to limit their initial capital investment and also to make practical assessments of the area's suitability for their purposes before longer or permanent commitments are made.

If a rural area is to attract manufacturing it must not only offer sites for new factories and make available existing empty buildings, but also provide extra housing for workers. Local authorities can encourage new forms of development by being ready to provide the necessary housing. It is unlikely that new firms will be able to find in the countryside all the skilled workers they require. Key workers will probably have to be "imported". Such mobility is a crucial element in diversifying the local employment base that is necessary for the development and future health of the countryside. Building houses and factories needs to be synchronized. One without the other creates problems and friction.

The size of catchment area from which a firm can draw its labour depends on its precise location, its situation in relation to other settlements, and the relative availability of public transport. Firms may run their own works' buses. Others may make a contribution to the cost of travel by their employees. Research in central Wales by H. D. Watts (1968) has shown that radii of up to 22 km define labour catchments in that particular region. Such a critical distance may not, however, apply in other rural areas with different terrain, settlement, and transport conditions. In many cases the bulk of the labour force in the factories of central Wales is composed of employees drawn from the surrounding countryside rather than from the town or large village in which the factory is actually located.

Generally, the local labour supplies available for industries establishing in rural areas have little or no industrial skill, and hence require training or retraining. This is not necessarily a difficulty, and some firms find it advantageous to train fresh labour whose outlook is not old fashioned or based on restrictive practices. In other cases the lack of industrial background and tradition can be a handicap. This disadvantage stems from the slowness of pace and lack of urgency that are frequently reported on the part of former agricultural workers.

The purpose of developing industry in mid Wales was to stimulate and diversify the local economy so that depopulation might be stemmed. Experience has shown that the scatter and sparsity of population in the region precludes the attraction of industry at the necessary speed and scale to stem depopulation. It is important to provide a range of employment in small country towns to retain a cross-section of the population, especially the better-educated and more able people. Enormous benefits, such as additional purchasing power and rate revenue, are eagerly anticipated from industrial installation, but some authorities tend to overlook the fact that retaining or attracting people imposes responsibilities for providing services, houses, roads, and schools. Industrial development in the countryside can prove to be a very costly business for a small authority.

Central Wales, like many rural areas, has a settlement pattern long since out of date. The Beacham report on depopulation in mid Wales (1964) commented that "it is probably true to say that the settlement pattern, combined with the low level of population, is a basic cause of many of the problems of mid-Wales" (p. 5). It also indicated that the area would benefit substantially if its scatter of population were reduced by nucleation into fewer, larger settlements. The Association has co-operated with county planning officers and government departments regarding the feasibility of expanding seven of the region's small towns. A small authority may have adequate powers to enable selected settlements to be expanded but it may not possess sufficient financial resources. In addition, small authorities may have difficulty in introducing objectivity and resolve in planning. "Future prosperity tends to succumb to present interests. Issues as to the line of a road, the use of certain areas of land, and even the possible introduction of people of a different culture, assume locally such enormous proportions that the objective of the town's development is lost sight of and is not pursued. These factors combine to raise again the need for regional agencies which, working within the framework of national policy, can make objective decisions as to a region's development and can have the powers and means to carry them out" (Garbett-Edwards, 1972a, p. 71).

The Mid-Wales Industrial Development Association does not have such powers. Its only weapon has been persuasion. Nevertheless, by providing fifty-three new factories and jobs for 4000 people it is helping to create a demand for people in an area which has known only decline for over a century. It has been easier to provide new jobs in light industry for women and youths than replacement jobs for middle-aged men leaving farming or quarrying work. The transition from primary to secondary or tertiary employment inevitably involves social upheaval and painful readjustment. By contrast, young people, who have not known any other form of employment, take far more readily to factory jobs which offer not only higher wages and better working conditions, but also convey social prestige.

Since 1968 the development association has been joined by the Mid-Wales Development Corporation. Both bodies have a joint administration and have the same chief executive officer. The task of the corporation is to expand Newtown, Montgomeryshire, by introducing new industries and doubling the resident population from 5500 to 11,000 in the 1980's. With its own powers, finance, and staff resources, the Corporation has tackled this task with determination. So far, not only has it completed all the statutory planning processes, but it has built and filled 150 houses, is building a further 300, and already has 350 more planned. It has built five factories (so far providing 300 jobs) and is building another four advance factories and seven nursery factories. It is constructing a £1.5 million sewerage scheme, and is soon to build a 40,000 square feet office block in the existing town centre. The population of Newtown has already risen to over 6100 (Garbett-Edwards, 1972b, p. 7). The other six "growth" towns of central Wales have so far been left largely to their own resources. Except in the field of industrial development, little special help has been available to them. Planning studies of each town have been undertaken and worthwhile small-scale development has been achieved in some cases. But left to their own insufficient human and financial resources, the authorities concerned have not on their own been able to implement effectively the strategy on which the area's expansion and future well-being must largely depend.

The various social and economic problems of remote rural regions, such as central Wales, need to be tackled by developmental and integrated planning policies to utilize the regions' various resources for forestry and tourism as well as for agriculture and industry.

REFERENCES AND FURTHER READING

The consequences of industrial installation in the countryside are considered in:

BRACEY, H. E. (1963) *Industry in the Countryside*, Faber, London.

BRACEY, H. E. (1970) *People and the Countryside*, Routledge & Kegan Paul, London (especially chapter 4).

GARBETT-EDWARDS, D. P. (1972a) The establishment of new industries, in ASHTON, J. and HARWOOD LONG, W. (eds.), *The Remoter Rural Areas of Britain*, Oliver & Boyd, Edinburgh, pp. 50–73.

GARBETT-EDWARDS, D. P. (1972b) The scope of rural life: development in mid-Wales. Mimeographed paper presented at the Town and Country Planning Association's Conference, London, 16–17 February 1972.

The management problems of central Wales are presented in:

BATEMAN, D. I. (1969) The future development of mid-Wales: the economic structure, *Journal of Agricultural Economics* **20,** 57–62.

DAVIES, M. L. (1967) The future of the small family farm, *Town and Country Planning* **35,** 497–502.

DAVIES, M. L. (1968) Mid-Wales: regional problems and policies, *Journal of the Town Planning Institute* **54,** 75–78.

GARBETT-EDWARDS, D. P. (1967) The development of mid-Wales: a new phase, *Town and Country Planning* **35,** 349–52.

GARBETT-EDWARDS, D. P. (1969) The means and pattern of growth, *Journal of Agricultural Economics* **20,** 63–67.

GRIFFITHS, E. G. (1969) Some aspects of the agriculture of mid-Wales, *Journal of Agricultural Economics* **20,** 69–79.

HILLING, J. (1968) Mid-Wales: a plan for the region, *Journal of the Town Planning Institute* **54,** 70–74.

HMSO (1964) *Depopulation in Mid-Wales*, London.

LEWIS, G. J. (1967) Commuting and the village in mid-Wales, *Geography* **52,** 294–304.

MOUNFIELD, P. R. and WATTS, H. D. (1968) Mid-Wales: prospects and policies for a problem area, in BOWEN, E. G. *et al.* (eds.), *Geography at Aberystwyth*, University of Wales Press, Cardiff, pp. 167–95.

ROWEY, G. (1970) Central places in rural Wales, *Tijdschrift voor Economische en Sociale Geografie* **61,** 32–40.

THOMAS, J. G. (1972) Population changes and the provision of services, in ASHTON, J. and HARWOOD LONG, W. (eds.), *The Remoter Rural Areas of Britain*, Oliver & Boyd, Edinburgh, pp. 91–106.

WATTS, H. D. (1968) Mid-Wales: the industrial journey to work in a rural area, *Journal of the Town Planning Institute* **54,** 79–80.

WELSH COUNCIL (1970) *Land-use Strategy: Pilot Report*, Cardiff.

Case studies of social changes taking place in rural areas following industrialization are provided by:

BERTRAND, A. L. and OSBORNE, H. W. (1960) Rural industrialization: a situational analysis, *Rural Sociology* **25,** 387–93.

BOER, J. (1966) Industrialization and urbanization in the province of Drenthe, the Netherlands, in HIGGS, J. (ed.), *People in the Countryside*, National Council of Social Service, London.

LUCEY, D. I. and KALDOR, D. R. (1969) *Rural Industrialization: the impact of industrialization on two rural communities in western Ireland*, Chapman, London.

CHAPTER 12

PASSENGER TRANSPORTATION IN RURAL BRITAIN

As IN other parts of the developed world, public transport services in rural parts of Britain have undergone important changes in the past two decades linked to contracting demands in response to the increasing ownership of private cars. Both railways and buses have been affected. Most rural services were no longer viable during the 1950's and many have since been withdrawn. Now, in the 1970's, the demand for public transport in the countryside may be small, but it stems from the elderly, the poorly paid, and other less-privileged members of society. Local train services have already disappeared from much of rural Britain. The following discussion will therefore concentrate on the contraction of rural bus services and the serious social problems which have to be faced when they are withdrawn.

RAIL SERVICES

Many rural railway lines have been closed because they carried too few passengers to be viable. Services on main lines which included stops at small country stations have been replaced by fast inter-city services which run direct between major concentrations of population. Important railway closures had taken place during the 1930's when the effects of the economic depression were being felt most badly. J. A. Patmore (1966) has shown that in 1930 and 1931 alone, passenger services were withdrawn from almost 800 km of line in England and Wales. This exceeded the overall total of railway closures that had taken place prior to that date. But since the nationalization of Britain's railway companies in 1948 many more sections of line have been closed. In particular the rate of closure speeded up after 1955 when the railways, for the first time, failed to balance even their operating account. Between 1950 and 1962 6750 km of track were closed throughout Britain. This figure compares with the closure proposals presented in *The Reshaping of British Railways* (the Beeching report) in 1963. In this report the British Railways Board showed that in 1962 40 per cent of the railway network had carried only 3 per cent of the total traffic (Patmore, 1965). In order to rationalize the situation the Board announced its intention to close almost 2500 stations and halts (55 per cent of the total) and to withdraw services from 8000 km of line (29 per cent of the 1962 total).

Rural areas suffered particularly harshly from the closures which resulted. Many branch lines had been built across the countryside in the latter half of the nineteenth century when the railway had become fully established and had achieved a virtual monopoly of surface transportation in Britain (Patmore, 1966). Such lines had been constructed to feed traffic on to the main lines. Many branch lines failed to yield an economic return even in their Victorian heyday. A sense of social obligation to rural residents on the part of railway operators was the sole reason which kept many of these rural lines open during the first half of the twentieth century when only the most blatantly unremunerative routes were closed. There were great variations in practice between the attitudes of individual railway companies. Most of the branch lines of the Great Western Railway in south-west England remained open. By contrast, many rural services operated by the London and North Eastern Railway in such areas as the Vale of York and East Anglia were closed. Perhaps the really surprising thing was the retention prior to World War II of so many rail services which failed to pay their way in rural Britain. As a result of the implementation of the 1963 proposals, most of Highland Britain is now without railway lines with the exception of main inter-city routes which pass through without stopping (White, 1963). Wide areas of countryside in Lowland Britain are also without services.

On a national scale trends in domestic passenger transportation have changed dramatically since 1952 (Table 22), with serious losses being registered by both railways and buses in contrast with massive increases in the volume of movement by car. Undoubtedly H. E. Bracey (1970) was correct when he noted that "the future of movement in the countryside lies with the private motor vehicle and the roads and not with the frequency of the trains or the bus services" (p. 75). Nevertheless, serious social problems have already arisen since a significant proportion of rural dwellers (including the young, the elderly, and the poorly paid) does not have access to its own personal means of transportation and, as a result, has to rely heavily on country buses. Recent withdrawals of bus services in the countryside have exacerbated their problems which will intensify as even more closures are implemented.

TABLE 22. NATIONAL TRENDS IN PAS-
SENGER TRANSPORT (SHARES OF TOTAL
PASSENGER MILEAGE)

	Rail (%)	Bus (%)	Car (%)
1952	21.5	44.7	33.8
1969*	8.9	14.7	75.9

* Air transport provides the remainder.

Source: DEPARTMENT OF THE ENVIRONMENT (1971b), *Study of Rural Transport in West Suffolk*, Report of the Steering Group, London, p. 3.

THE RISE AND FALL OF RURAL BUS SERVICES

The disparity between what may be considered socially desirable and what is economically viable is well illustrated by the current controversy surrounding the provision of bus services in rural Britain. The foundations of the rural bus network in Britain were established during the 1920's to take over the transportation role that had previously been played by carriers' carts which moved goods and people from their home villages to the nearest railway station (HMSO, 1961). After 1918 many of these carriers' services were developed into motor-bus services. Ex-servicemen had learned to drive during the war and used their gratuities to start bus and garage businesses. At the same time, larger companies were establishing main-road services. These were in competition with the railways which began to lose local passenger traffic to the cheaper, more flexible and more convenient country bus routes. A licensing system was introduced, based on vehicles rather than services, so that once licensed a bus could be operated freely without restriction regarding time, place, or fares. The results were chaotic. In 1930 a new licensing system was introduced with licences being granted for services on specified routes.

During the 1930's the number of buses and bus operators in Britain declined by one-fifth, but the volume of passenger journeys and vehicle kilometres continued to increase. Immediately before World War II bus operation was still a small-scale activity with 4000 of the 4700 operators running five vehicles or less. The bus industry continued to expand during the war and also in the immediate post-war years when petrol was rationed and the manufacture of private cars was limited. In 1950 Britain had a more extensive bus system and a much better standard of public road transport than during the 1930's. A high level of demand persisted for a further 5 years, but in 1955–6 the number of passenger journeys on bus routes began to decline. Between 1955 and 1959 the number of passenger journeys fell by 10.2 per cent by contrast with the growth rate of 10.0 per cent in the years 1948–51 and 2.3 per cent in the 4 years 1951–5.

This contraction in overall demand affected rural bus routes earlier and more acutely than town services which, however, were also to experience contraction soon. As demand declined, many rural routes were no longer viable. Some operators cross-subsidized rural routes with profits from urban buses. However, the total number of rural services was reduced and a vicious downward spiral of service contraction was started. Operators raised fares to try to cut their losses. As a result, fewer passengers used the services and even smaller revenues were produced. Services were cut again, but costs continued to rise and fares were raised once more. Because rural bus services were both costly and by now generally infrequent, ever-rising numbers of country dwellers purchased their own cars. The decline in passenger numbers using buses in fact far outstripped reductions in services. Between 1952 and 1969 the number of passengers carried on ordinary time-tabled bus services in both rural and urban areas of England and Wales fell by three-eighths but over the same period the mileage of bus services operated fell by only 13 per cent.

Figure 12.1 depicts the pattern of bus services operated by the regional contractor in north-western Norfolk over the same period. In fact, the network remained substantially similar with few services being withdrawn and settlements thereby losing their bus connections. Nevertheless, important reductions in services took place on some routes, largely in response to the closure of large air force bases which had required a large civilian personnel for their operation in the 1950's and early 1960's. Of necessity, such a labour force commuted to the bases from urban centres in the area and very often travelled by public-service bus. Some new routes were added during this period in order to guarantee passenger transport along routes that had previously been served by the railway before all services were withdrawn from this part of East Anglia. The apparent stability of the network, which emerges from a comparison of the two maps, should not detract from changes in service frequency. The associated problems of declining passenger numbers and falling revenues affected north-west Norfolk as well as all other rural areas over this period. Dramatic changes may be expected in the immediate future as decisions are taken regarding which routes should continue to operate with local authority and Exchequer subsidies and which should be withdrawn.

DECLINING DEMAND FOR RURAL BUSES

The growth in private transportation both contributed to and resulted from such a downward spiral of public service provision. In Britain increasing car ownership had been delayed in the immediate post-war years, and not until 1949 was the number of cars existing in 1939 regained. Thereafter the number of private cars in Great Britain increased rapidly from 2.5 million vehicles in 1952 to 11.2 millions in 1969. At present there are 21 cars for every 100 people in Britain. Estimates by the National Road Research Laboratory suggest that a rapid rise will bring the ratio to 28:100 people by 1975 (Devon Report, 1971). Over the same period car ownership in the countryside will rise to an above-average level, for example from 28:100 to 35:100 in rural parts of Devon. The sample census of 1966 has already shown that 46 per cent of all households in England and Wales owned one car or more, with higher proportions (over 55 per cent) being registered in rural areas than in towns.

The rise in private-car ownership triggered off the downward spiral. But, at the same time, the curtailment of road and rail services stimulated further purchase of private transport, even among lower income groups whose members might otherwise not have chosen to do so. The growing practice of giving lifts in rural areas was partly a response to inadequate transport services, but at the same time aggravated conditions even more. The receipts of bus operators continued to fall during the 1960's not only in rural areas but in towns as well. Opportunities for cross-subsidization became even less.

Changes in entertainment habits following the spread of television also had their effect on demand for public transport. The number of TV sets increased fivefold between 1952 and 1960. Cinema attendances fell by one-half over the same period. Such swift changes accentuated the problems of bus operators by

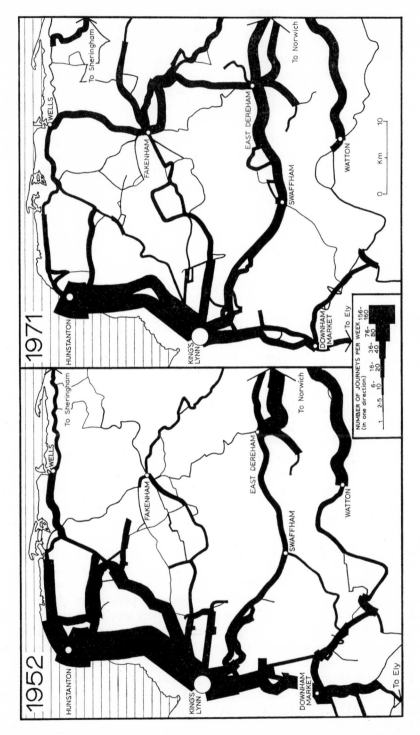

Fig. 12.1. Bus services operated by the regional contractor in north-western Norfolk, 1952 and 1971.

greatly reducing the volume of demand for public transport on Saturday after-noons, in the evening, and other off-peak periods. Modifications in the distri-bution of population in the countryside caused further difficulties as residents tended to move to larger villages and country towns. The remaining population in hamlets and villages was reduced even further, and demand fell on bus routes serving smaller settlements.

Some observers claimed that increasing tendencies for groceries and other goods to be delivered from country towns and for mobile shops to visit the countryside reduced the demand for rural buses for shopping trips. Nevertheless, the Jack Committee (HMSO, 1961) reported: "We have no convincing evidence of the effects of mobile shops and delivery vans on bus travel for shopping purposes but we think that the effect is not very great as regards rural bus services. We think that most country housewives still prefer to visit their market town at least once a week, partly because of the attraction of the outing, and partly because of the wider range of commodities to which they can have access and the price com-parisons they can make" (p. 9).

FINDINGS OF THE JACK COMMITTEE

The precise contributory causes for the decline of rural bus services remained a subject for debate, but it was clear that public transportation in rural Britain was contracting fast in the second half of the 1950's. As has already been discussed, uneconomic rail services were axed, and for a while buses provided cheaper and more flexible replacements. "It might be expected that where an uneconomic rail service is withdrawn the substitute bus service would be better able to make a profit or at least to incur a smaller loss. This, however, has not always happened and we have encountered cases where substitute bus services, after an initial period, have also been withdrawn" (HMSO, 1961, p. 9). Losses continued to be made by the bus operators as wages rose and as fuel tax was increased. Such a situation led to the creation of a special committee in 1959 under the chairmanship of Professor D. T. Jack "to review present trends in rural bus services and in particular to enquire into the adequacy of those services; to consider possible methods of ensuring adequate services in the future; and to make recommendations".

The Committee's report contained valuable information on rural public transportation in other countries. Public services in the United States had declined much further than in Britain, not only because of the sparse distribution of popu-lation in many country areas but also because of the very high percentage of car ownership. As early as 1954 one-quarter of all communities in New England with over a 1000 residents were entirely without public transport. In the same year 88 per cent of all passenger kilometres throughout the United States were travelled by car and only 4 per cent by bus. Table 22 shows that Britain is still far from reaching such a situation.

Conditions in Sweden were similar to those in Britain. A special transport commission had advised the Swedish Government and had recommended raising rural bus fares and integrating the operation of schools' buses with adult movements. In cases where demand was very slight, the commission envisaged that the licensed operation of private cars might be a better answer than the provision of buses.

In other European countries, such as Switzerland and West Germany, rural buses were not losing passengers during the 1950's even though rates of private car ownership were rising. In both countries rural transport provision was integrated with postal service movements. The West German post office was required to maintain bus services in thinly populated areas as part of the Government's social policy. In Switzerland the post office was the traditional operator of rural buses and co-ordinated its time-tables with those of the railways. Fares on Swiss buses were closer to the economic rate than those charged in Great Britain, Sweden, or West Germany. The Jack Committee was particularly interested in Swiss techniques for combining bus operation with postal deliveries; for determining the adequacy of services; co-ordinating the time-tabling of bus and train movements in rural areas; and using sub-contractors to operate some services.

Before making recommendations, the Committee sought to discover which sections of rural communities suffered when country bus services were reduced or withdrawn. The following groups were identified: young people attending further education classes and seeking entertainment in the evenings; commuters from rural areas to jobs in town; elderly folk who might have to travel to post offices in neighbouring settlements to draw pensions; housewives needing to shop in town for other than day-to-day necessities; patients visiting doctors' surgeries and requiring prescriptions to be made up; visitors to patients in hospitals and out-patients themselves attending for treatment.

The Committee reported on a number of ways for rationalizing public transport. The possibility of using school buses for adult as well as child passengers was raised. The Committee thought that this could make only a slight contribution since school buses were completely filled with children at the times of the day which corresponded with journeys to and from places of work and shopping centres.

Experience in the use of minibuses was reviewed. The Committee was not greatly in favour, since a minibus was only marginally cheaper to run than a single-decker bus. In both instances the greater part of costs was made up by drivers' wages. Minibuses could not cope with occasional demands for handling large loads.

In 1956 the Ministry of Transport explained how the carriage of goods and passengers by a modern version of the village carrier might be achieved in areas which could not support a regular bus service. It was suggested that local tradesmen travelling regularly by van to the nearest town might apply for a licence to carry passengers. This proposal was not widely supported since additional insurance would be required as well as licences. Many villages which were too small for a public bus were also too small to have a suitable carrier.

The idea of the postal bus, combining the carriage of passengers and mail, raised serious problems in Britain even though it worked successfully in parts of Europe. The complications which arose were linked to differences in timing and delivery practices. In Switzerland and West Germany, postal buses took mails to sub-post-offices whence they are delivered by postmen on foot or cycle. In Britain, post-office vans deliver to outlying farms and thus pursue devious routes that would be impractical for postal buses.

Further fare increases were considered by the Committee as making the rural bus service problem worse rather than better. The Committee believed that adequate rural transport could only be provided with a measure of financial assistance from outside the industry. A reduction or remission of fuel tax on rural services would provide only a short-term solution. Direct financial aid was far more desirable. The cost would be borne partly by the Exchequer and partly by the various county councils involved. The tasks of determining which routes were important enough to merit the receipt of subsidies and actually administering the financing system should be handled by the county councils.

These proposals were embodied in the 1968 Transport Act which empowered local authorities to give financial aid at their own discretion to improve or maintain passenger transportation in the countryside. Exchequer grants would cover half of the approved subsidies. On 1 January 1971 local councils were given a full year to decide which rural bus services should be kept on with subsidies and which should be terminated. The existing system of rural services soon received a serious jolt. Local county councils reacted in different ways, but a flood of three-month closure notices was announced in many parts of the country.

WHO USES RURAL BUSES?

Detailed factual investigations had been undertaken in six rural sample areas (parts of Devon, Montgomeryshire, Lincolnshire, Westmorland, Kirkcudbrightshire, and Banffshire) in the spring of 1963 following the Jack Committee's report. These plotted the actual use made by members of 6000 rural households of various forms of transport (HMSO, 1965). Over one-half had the regular use of some private means of transport. More than 40 per cent of households had their own private car. This was well above the national average of one-third at that time. Only 10 per cent of the total journeys made during the investigation period involved public transport—essentially by bus since only one of the six areas still had a train service. Sixty-five per cent of journeys were by private car or van, 10 per cent by contract buses (for schoolchildren and factory workers), and 15 per cent by other means, such as bicycles, walking, taxis, or hired cars.

The overall use of public transport was slight, but people used it for making one-fifth of all shopping journeys in three of the sample areas. Other forms of journey were not greatly dependent on service buses, which carried only 10 per cent of journeys to school (contract buses carrying over 50 per cent of such movements) and 10 per cent of journeys to relatives and friends. Proportions of journeys to work by public bus varied from less than 5 per cent to 25 per cent according to the sample area. Buses did play a very significant role in journeys for medical and dental treatment. Members of the 6000 households investigated in 1963 were asked if they were in any way restricted by the existing level of transport provision. Between one-half and three-quarters of households without their own cars said that they were hindered, mainly with regard to obtaining medical or dental treatment. Only one-fifth of car-owning households said that they were hindered.

An investigation of mobility patterns of residents in five parishes in north Norfolk in June 1970 showed the very limited use made of buses for any purpose

(Munton and Clout, 1971). Only one individual out of the 306 cases investigated travelled to work by public bus. Employment opportunities in each of the five parishes were very limited, and hence those who were geographically immobile tended to be occupationally immobile as well. The situation was particularly serious for those on low wages with little prospect of running a car. Agricultural workers, whose chances of continuing in farm-based employment contract each year, most clearly typify this problem. However, the greater proportion work locally (84.7 per cent in their home parish) and travel to work on foot or by bicycle (Table 23). Their mode of travel was very different from that of the remainder of the total work force.

Each sample parish was served by travelling shops on which many elderly, housebound people relied entirely. In addition, each of the five parishes had at least one general store that catered for its inhabitants' normal day-to-day needs. The primary significance of poor bus services in shopping behaviour therefore related to the wishes of rural residents to shop in nearby towns where greater ranges of goods were offered than in their home villages and at lower, supermarket, prices. Many residents complained that high bus fares offset any economic saving from shopping in town unless large quantities of provisions were purchased each time. An analysis of shopping habits showed that, first, the great majority of households with cars visited local towns for their shopping requirements, whilst households without cars made little or no use of urban shopping facilities. Second, not only was the frequency of visits to urban shops much greater among car-owning households, but the number of shopping centres visited was higher, which permitted a wider range of shopping experience and a greater degree of purchasing choice. Only 5 per cent of the total households investigated used the bus to go shopping on any occasion. Existing services were not only inadequate for that purpose but were also considered to be too highly priced.

The bus services provided by the regional contractor in this part of Norfolk were irrelevant to journey-to-work behaviour and were of only limited importance in shopping movements. Further cuts in services would in most cases make little difference to mobility patterns. However, there were sections of the community which would suffer from any reductions, namely the aged and the less affluent who had to rely on public services for choice in their shopping activities and access to doctors, dentists and other town-based services. The analysis showed that rural transportation is not just an interim problem as is sometimes supposed. Undoubtedly, the proportion of carless families will decrease in the future, but it would be unrealistic to assume that the demand for public transport services from the poor, the elderly, and the sick will disappear completely.

Similar results emerged from detailed studies of the use of rural transport by 6000 people in Devon (fifteen parishes) and west Suffolk (nine parishes) commissioned by the Department of the Environment in Autumn 1970. Only 6 per cent of the sample made regular and frequent use of public transport, that is every day or most days. Twenty-four per cent used public transport more frequently than once a month, but 55 per cent never or hardly ever made use of it. Tables 24 and 25 show the actual use of various forms of transport for specific purposes. Public transport of all kinds accounted for between 16 per cent (west Suffolk) and 22

TABLE 23. NORFOLK STUDY: PLACE OF WORK AND MODE OF TRAVEL (%)

	Cases	Place of work		Mode of travel						
		Inside home parish	Outside home parish	Own car	Private lift	Works bus	Public bus	Bike	Foot	Other*
Total work force	306	46.1	53.9	37.9	4.7	12.3	0.3	21.3	13.6	9.9
Agricultural workers	59	84.7	15.3	11.9	0.0	6.8	0.0	50.8	25.4	5.1
Remainder of work force	247	36.8	63.2	43.3	5.7	14.4	0.4	13.8	10.5	11.9

* Includes living at place of work.

Source: MUNTON, R. J. C. and CLOUT, H. D. (1971), The problem bus, *Town and Country Planning* **39**, 115.

TABLE 24. DEVON: RELATIONSHIP BETWEEN MODE OF TRANSPORT AND JOURNEY PURPOSE (%)

	All trips	Work	In course of work	School	Shopping	Personal business	Leisure	Taking other people
Bus or coach: Local	6	7	*	8	13	5	5	*
Works	2	6	2	*	—	—	—	*
School	13	*	—	63	*	—	*	*
Train	*	*	*	1	*	*	*	—
Taxi or hire car	*	*	—	*	*	1	1	*
Total public	22	14	2	78	14	7	8	*
Car or van: Household	61	58	56	16	75	79	75	91
Firm's	5	9	23	*	1	1	2	3
School	*	*	—	4	—	*	*	*
Borrowed/arranged lift	8	10	2	4	5	9	11	1
Lorry	*	1	3	*	*	*	*	—
Motor cycle, scooter	*	2	—	*	*	*	*	—
Cycle	2	6	*	1	1	*	1	*
Walked	2	2	*	2	2	2	2	*

* Less than 0.5% (percentages rounded up).

Source: DEPARTMENT OF THE ENVIRONMENT (1971a), *Study of Rural Transport in Devon*, Report of the Steering Group, London, p. 24.

per cent (Devon) of all journeys.* Bus services provided by regional contractors and local operators accounted for only 6 per cent of journeys in both areas. This may be compared with 10 per cent in the 1963 surveys. Apart from shopping activities, rural buses were of minimal importance in passenger movement in rural parts of west Suffolk and Devon in winter 1970/1. Schools' buses catered for three-fifths of all educational journeys. The private car was unquestionably the most important means of transport for every purpose except the journey to school.

Great variations in the rates of use of rural buses emerged between different sections of the population. As other studies had shown, it was young people, housewives, and the elderly who were most likely to be without their own forms of transport and therefore depended on public services. They would suffer most if bus services were to be withdrawn. Already one-third of the respondents to the 1970/1 surveys replied that they were unable to make journeys they would have liked to undertake due to an absence of suitable public transport. This often involved leisure activities and particularly affected young teenagers who had no vehicles of their own.

In the words of the West Suffolk report (1971) the poor provision of rural transport "at best represents a sad limitation of [the residents'] ability to lead a full life, at worst it represents substantial hardship. . . . The real trouble in rural areas . . . seems to be that these 'residual' needs for transport are often so scattered and varied that it is difficult to envisage their being matched together to form any form of sensible public transport load" (p. 12). It was considered that the situation would deteriorate rather than improve, as education and health services and facilities for shopping and entertainment became concentrated at fewer key settlements in the countryside.

Large numbers of rural car owners gave lifts to take neighbours to shops and to other town-based facilities. Only a quarter of the car-users interviewed in Devon and west Suffolk reported that they never did so. The pride of some people, especially the elderly, would not allow them to ask for help even when it would be freely and willingly given. The reports considered that some of the present limitations of lift-giving might be overcome if car drivers were allowed by law to receive payment for lifts. People would then perhaps be less reluctant to ask since they would no longer appear to be seeking charity. Many drivers might see an opportunity for the reimbursement of some of their expenses and would be more ready to offer lifts on a prearranged basis.

RURAL TRANSPORT IN THE FUTURE

In the long term, the whole *raison d'être* of rural transport services will have to be re-thought and special attention be paid to the precise needs of those who actually use rural buses. The Department of the Environment's reports and public reactions to the spate of closures which took place in 1971 (and the even larger number which threatened) have thrown some light on the issue. Mr. John Gilkes, Secretary of the Rural District Councils' Association, noted that many

---

* Note that school buses and works buses were excluded from the "public transport" category in the 1963 surveys quoted previously.

TABLE 25. WEST SUFFOLK: RELATIONSHIP BETWEEN MODE OF TRANSPORT AND JOURNEY PURPOSE (%)

	All trips	Work	In course of work	School	Shopping	Personal business	Leisure	Taking other people
Bus or coach: Local	6	6	*	3	19	7	3	*
Works	1	3	*	*	*	—	*	—
School	8	*	*	57	*	*	*	—
Train	*	*	1	*	*	2	*	*
Taxi or hire car	*	*	*	*	*	2	1	*
Total public	16	10	2	62	20	10	5	*
Car or van: Household	57	54	55	17	63	65	70	85
Firm's	9	12	34	2	5	7	5	9
School	*	*	—	4	—	*	*	
Borrowed/arranged lift	7	7	2	5	6	11	12	2
Lorry	*	*	3	—	*	*	*	—
Motor cycle, scooter	3	6	*	*	*	1	2	*
Cycle	5	9	1	5	3	3	3	1
Walked	4	2	1	11	2	2	3	2

* Less than 0.5% (percentages rounded up).
Source: DEPARTMENT OF THE ENVIRONMENT (1971b) *Study of Rural Transport in West Suffolk*, Report of the Steering Group, London, p. 29.

country people would find themselves worse off for transport than their ancestors were 100 years ago. In addition to restricting the mobility of the elderly and the poor in remote parts of the countryside, reductions in bus services are also affecting families living in villages close to large cities, typically young married couples with either a mortgage or a car or both. Wives and young children in one-car families are marooned during the day if the husbands take the car to drive to work.

Rural dwellers in Britain have had to look hard at the various possibilities for providing alternative forms of transport in the countryside.

(i) Schools' buses operated under contract with local education authorities are used intensively twice a day for 5 days a week during the school terms. The buses are filled to capacity for schools' journeys in the early morning and late afternoon, but it has been suggested that they might be used to take adult passengers in the middle of the day if time-tabling would allow.

(ii) Services using minibuses, maxi-taxis or hire cars have been arranged in some parts of the country. If they really were operated on a taxi basis one would have to telephone to summon them. The problem here is that elderly and poorer members of rural communities rarely have telephones or, indeed, have sufficient cash to pay normal taxi rates. By contrast, minibus clubs or car-pooling associations seem to offer far more reasonable solutions. The Devon and west Suffolk studies showed that where bus services had become very infrequent or had ceased altogether, people did adapt to the situation and many were able to get lifts. Local self-help already happened and could become more widespread. But good neighbourliness might not be sufficient in all situations. Arrangements might be co-ordinated by local groups (parish councils, voluntary organizations) so that those who were willing to offer lifts could be brought into contact with those who needed them.

(iii) Converting postal vans to postal minibuses which could take passengers has posed considerable problems in Britain, even though postal and passenger services have been successfully integrated in parts of Continental Europe. The Devon report (1971) showed that "in general the operational requirements of the post office do not match the transport requirements of potential passengers (the routes are circuitous, and the journeys at the wrong time of the day, moving outwards from towns in the morning, often with no provision for a return journey)" (p. 13). Postal minibuses, if they were to operate, would have to be carefully routed to ensure that journeys were not too lengthy and time-consuming between the most remote point served and the nearest town. This could only be achieved by eliminating journeys to remote farmhouses and by delivering only to mail boxes on the main roads or relying on sub-post offices for local deliveries, as in central Europe. It would be necessary to guarantee a round journey on those days when the service operated. This might involve a postal service operating twice on every other day rather than once every day. Other problems stem from the general issue of maintaining security and the fact that the present generation of post-office vans is not suitable for conversion to take passengers.

(iv) A fourth suggestion has looked at the problem in the long-term and involved the future construction of council houses and old peoples' accommodation only at chosen settlements along selected routes where public transport services would be guaranteed (T. Bendixson, 1971). Such a proposal would fit in well with policies for settlement rationalization in the form of key villages but would only be feasible in the long term. The elderly and those on low incomes are at present the least geographically mobile, being attached to the farms where they work or to the houses where they have lived most of their lives. Other forms of transport would be needed to resolve their immediate needs.

Short-term approaches to rural transport problems are developing along lines which offer greater flexibility. In July 1971 the Minister for Transport Industries announced new proposals regarding the road service licensing system which would help rural areas. The existing system originated at a time of rapid expansion of the bus industry and has contributed largely to the present route system. With the decline of bus patronage there is no longer a clear case for such rigid controls, especially in rural areas where the emphasis must now increasingly be on informal arrangements, often amounting to no more than a lift in a private vehicle for payment. Under the Minister's proposals, cars and minibuses would be exempt from road service licensing, together with tours, excursions, and schools' buses. No public service vehicle seating fewer than eight people would require a special PSV licence. When these proposals become law, car owners will be permitted by law to give lifts for payment and operators will be able to start minibus services without tussles about the route with local bus operators. Reactions to these suggestions have been mixed. Some rural dwellers have seen them as useful ways of helping the car-less. Bus companies have viewed them as a further threat to the remaining rural services which local authorities have agreed to subsidize. In October 1971 the Minister announced further measures to try to ease the situation such as a one-third increase in grants to the bus industry for fuel tax rebates and for the purchase of new vehicles.

These measures fit in with the threefold proposals for rural transportation in the future which emerged from the 1970/1 studies. These urged, first, that small and irregular movements of rural residents should be catered for by cars and small vehicles, with a relaxation of the licensing system. Second, a variety of types of vehicle should operate on rural routes which were more frequently used, to include normal buses (if the volume of demand were great enough) but also minibuses and cars to deal with smaller but still regular demands. Finally, rural transport services should be co-ordinated and kept under constant review by local authorities. As the west Suffolk report (1971) succinctly put it: "the main problem in many areas is ceasing to be one of public transport. Rather it is one of catering for a residuum consisting of the needs of a small minority of the population, too dispersed any longer to justify conventional bus services, but with very real needs none the less" (p. 19).

REFERENCES AND FURTHER READING

The general theme of rural transportation in both an historical and a contemporary context is presented in:

BRACEY, H. E. (1970) *People and the Countryside*, Routledge & Kegan Paul, London (especially chapter 5).

The contraction of the railway network is discussed in:

PATMORE, J. A. (1965) The British railway network in the Beeching era, *Economic Geography* **41,** 71–81.
PATMORE, J. A. (1966) The contraction of the network of railway passenger services in England and Wales, 1836–1962, *Transactions of the Institute of British Geographers* **38,** 104–18.
WHITE, H. P. (1963) The reshaping of British Railways, *Geography* **48,** 335–7.

The evolution and recent problems of rural bus services are described in:

HMSO (1961) *Rural Bus Services*, Report of the Committee [the Jack Committee], London.

Local studies of the use of public transport in the countryside are contained in:

DEPARTMENT OF THE ENVIRONMENT (1971a) *Study of Rural Transport in Devon*, Report of the Steering Group, London.
DEPARTMENT OF THE ENVIRONMENT (1971b) *Study of Rural Transport in West Suffolk*, Report of the Steering Group, London.
HMSO (1963) *Rural Transport Surveys*, Report of Preliminary Results, London.
HMSO (1965) *Rural Bus Services*, Report of Local Inquiries, London.
NORTHERN ECONOMIC PLANNING COUNCIL (1972) *Rural Transport and Proposed Changes in the Bus Licensing System*, Newcastle-upon-Tyne.
ST. JOHN THOMAS, D. (1963) *The Rural Transport Problem*, Routledge & Kegan Paul, London.

The distribution of disused railways and the recreational and other uses to which they may be put are discussed in:

APPLETON, J. H. (1970) *Disused Railways in the Countryside of England and Wales*, HMSO, London·
ANON. (1970) *Schemes for the Recreational Use of Disused Railways*, Countryside Commission, London·

Investigations of rural transportation by geographers include:

DICKINSON, G. C. (1960–1) Buses and people: population distribution and bus services in East Yorkshire, *Town Planning Review* **31,** 301–14.
JOHNSTON, R. J. (1966) An index of accessibility and its use in the study of bus services and settlement patterns, *Tijdschrift voor Economische en Sociale Geografie* **57,** 33–37.
MUNTON, R. J. C. and CLOUT, H. D. (1971) The problem bus, *Town and Country Planning* **39,** 112–16.

Proposals for replanning rural settlements in relation to public transport routes are outlined by

BENDIXSON, T. (1971) Keeping on the country buses, *New Society*, 18 February 1971, p. 275.

CHAPTER 13

INTEGRATED MANAGEMENT OF THE COUNTRYSIDE

THE diversity of recent social and economic changes in the countryside has been outlined in preceding chapters. For ease of presentation a number of issues were isolated for consideration in individual sections. However, three main reasons make such a systematic treatment something of an unreal exercise. First, all of the component themes are interlinked in a dynamic system, so that changes in one component will have repercussions on all the others. By way of simple illustration, a reduction in the total number of residents in a given area of countryside will have great implications for service provision, public transport, the possibility of structural changes in farming, the chance of houses becoming available for acquisition as second homes, and so on.

The second point to be raised is that whilst all component themes are linked together in functional terms, the particular combination will vary between rural regions. At a high level of generalization one might be able to depict common management problems in all rural areas, but clearly there are differences in degree between the issues encountered, e.g. in the Highlands and Islands of Scotland and those in East Anglia.

Third, there are important variations in the rate of social and economic change between regions, just as there are different mechanisms for management, that is for steering change, in the planning frameworks of individual countries. All these reasons point to a need to consciously adopt an integrated approach to management problems in the countryside, tackling issues on a spatial or geographical basis. Such an approach would allow broad policies to be modified to take account of the specific social and economic trends of particular areas and also to consider the varying potential of their physico-geographical environment.

Integrated management proposals have been implemented by rural development corporations operating in parts of France and other regions of western Europe. Such corporations in the Languedoc, the Massif Central, and elsewhere recognize not only recent trends in population change and agricultural production but also the need to diversify the local economy through the development of manufacturing, afforestation, and the recreation industry. Their work provides good examples of "applied geography", in the complete sense of the term, since spatial variations in land capability, microclimate, slope, and other aspects of physical geography are investigated while plans are being prepared, in addition to attention being paid to demographic, social, and economic parameters. Rather less has been achieved in Great Britain than in France for drawing up integrated management proposals in the countryside. Two main reasons contribute to this relative lack of success.

First, until very recently physical planning in Britain was very city-orientated. In A. S. Travis's (1972) words: "The emphasis in the 1947 [Town and Country Planning] Act was urban even where the planning of counties was concerned. For instance, village planning was seen as town planning writ small and other policies such as for green belts, scenic conservation, screening of mineral workings . . . siting of pylons, were all part of an urban-based fitting of amenity and conservation into the countryside. It is significant that the use of land for agriculture and forestry was not deemed 'development' for the purposes of the Act and therefore the vast majority of the land resource of the rural counties was not subject to planning control" (p. 187). Second, a large number of authorities were concerned with management proposals for the countryside and hence there was little chance of satisfactory integration.

D. G. Robinson (1972) has outlined the need for a serious strategic overview of the policies of various types of agency as they affect the use of rural areas in Britain. Agencies concerned with the following themes would need consideration: (i) strategic economic planning (the various regional economic planning councils with no executive powers, and the Highlands and Islands Development Board with executive powers); (ii) planning public investment programmes for rural areas (Ministry of Agriculture, Forestry Commission, British Rail, etc.); (iii) planning for social, economic, and land-use adjustment (the short-lived rural development boards, the Crofters' Commission); (iv) physical planning (local planning authorities, including National Parks' boards and the Countryside Commission); (v) nature conservation (Nature Conservancy); and (vi) major private investment programmes, involving the projects of country landowners, various tourism developments, industrial installation, mineral exploitation, and other forms of development.

At present, no single agency is empowered to tackle the whole spectrum of management issues encountered in rural regions, ranging from aspects of economic production—through settlement and service provision—to amenity. The Highlands and Islands Development Board in north-west Scotland has powers which enable it to go some of the way towards meeting such a need for integration and allowing the planning agency to act as an effective catalyst for regional development. Attention will therefore be drawn to the aims and achievements of the HIDB later in this chapter. The proposal in the 1967 Agriculture Act for setting up rural development boards in England and Wales also offered the chance of integrating the management of key issues on a spatial basis. Unfortunately, the concept of the rural development board was rejected by the Conservative Government in January 1971. Nevertheless, the general objectives of the boards and the short life of that which operated in the northern Pennines will be discussed later.

The Planning Advisory Group, which was set up to evaluate the planning system established in England and Wales after the 1947 Town and Country Planning

Act, reported a "general impression that country planning . . . has tended to become a neglected aspect of planning work and the present development system tends to discourage a more positive approach".

A. S. Travis (1972) has insisted that "integration of land use is still a primary national requirement, but in this context integration means coordination of multiple use of land and associated industries within a given area of land. We do not yet have the rural land development authorities and the long awaited Ministry of Land and Natural Resources has come and gone, and therefore a close link between agricultural planning, national-resource planning and conservation, and general physical planning has yet to be achieved. The Study Group recommended that a new executive body such as a regional development authority should be created which would operate regionally and be responsible for local interpretation of central government policy. It should have the necessary power to coordinate effectively all forms of rural land use" (pp. 198–9).

Such an executive body would seem to offer a very good means of tackling land-use problems which need to be viewed at a variety of scales, namely nationally, regionally, and locally. It is also necessary to view town and country as integral components in broad functional regions so that the implications of the urbanization of the countryside, in all its forms, may be recognized. In addition, both social and economic issues need to be considered seriously in the spatial management of areas visually recognizable as "countryside".

D. G. Robinson (1972) outlined the type of agency that should be set up for the comprehensive and developmental management of the countryside component. He argued that "narrow traditionalism and the perpetuation of anachronistic features of the economy and settlement pattern are inimical to the developmental approach" (p. 215). However, once adopted, such an approach "should not ride roughshod over the culture and human characteristics of an area but rather enlist and encourage regional consciousness so that it becomes a dynamic and distinctive force for self help, growth and development, and yet also provides a safeguard for sympathetic treatment of the area's human and land resources" (p. 215). Any comprehensive agency should: (i) operate on a regional scale; (ii) treat economic, social, and physical planning as a combined operation; (iii) be innovative in outlook; (iv) have powers to intervene (namely executive powers and its own budget); (v) have a continuing role to play and be allowed to take a long-term view of problem solving; and (vi) be an indigenous agency involving local inhabitants, thereby avoiding alienation, misinformation, and misinterpretation which can result when management proposals are directed from central government. The agency should first draw up an inventory of its region's resources and estimate present and likely future demands on these resources. It should establish a system for undertaking continuing research and intelligence once its proposals are drawn up and also incorporate a mechanism for monitoring and reviewing its plans with the passage of time.

The only serious British attempt to achieve such an integrated approach is found in the Highlands and Islands Development Board. However in the late 1960's it seemed possible that proposals for rural development boards in parts of England and Wales would offer similar but not identical opportunities.

THE HIGHLANDS AND ISLANDS DEVELOPMENT BOARD

The 3.6 million ha of highlands and islands in north-west Scotland (Fig. 13.1) cover one-half of the total area of Scotland and one-sixth of that of Great Britain. The region has experienced serious depopulation and now contains only 276,000 inhabitants, 5 per cent of the Scottish population. Local communications are difficult and, in the context of Great Britain, environmental conditions are harsh. Arable land and improved grass cover only 240,000 ha. Many aspects of agricultural production are in need of modernization. Woodland covers 218,000 ha, but by far the greater part of the region is covered by rough grazing land. Social and economic problems in north-west Scotland have been investigated in several generations of official report from the Napier Commission report in 1884 to that of the Advisory Panel for the Highlands and Islands, which was set up in 1947 and was not dissolved until 1965. It recommended that there should be radical changes in the arrangements for controlling land affairs in the Highlands and Islands.

The result was the creation of the Highlands and Islands Development Board (1965) with greater powers than any previous type of organization that had been set up in the region. The Board has the dual aim of making further provision for the economic and social development of the region. Its jurisdiction operates over the whole of the crofting counties, thereby bypassing the administrative fragmentation of the region. It is empowered to review all matters affecting the economic and social welfare of the Highlands and Islands and is then required to submit proposals to the Secretary of State for Scotland. It order to tackle its far-reaching responsibilities the HIDB has been given suitably broad powers: to acquire land (even compulsorily if need be); to construct buildings; to provide equipment and services; to perform or to act as agent for any business development that will improve the region; and to provide advisory, training, management, accountancy, and other services to anyone wishing to engage in business or promote publicity in the Highlands and Islands.

Many schemes and projects for development in north-west Scotland failed in the past because of an inability to raise the necessary capital. If developers are able to show that their schemes will be worthwhile they can now receive the full backing of the HIDB to argue their case for grants, loans, and any other form of financial aid from the Secretary of State for Scotland, the Department of Agriculture and Fisheries, the Treasury, or whichever other body might be appropriate.

The first chairman of the HIDB, Professor Robert Grieve, summarized its role as tackling "areas that the various revolutions in agriculture, industry and technology have passed by—and to pull them on to their feet" (*First Annual Report*, 1967, p. 1). Thus the Board has adopted a multiple approach to the management of regional life involving farming, forestry, tourism, and manufacturing industry.

The HIDB argues that primary activities will remain at the base of the region's economy. The agricultural labour force declined rapidly, from 6100 to 4250 (—30 per cent) between 1966 and 1970, at a much faster rate than that recorded for Scotland as a whole. Much has been done by the Department of Agriculture

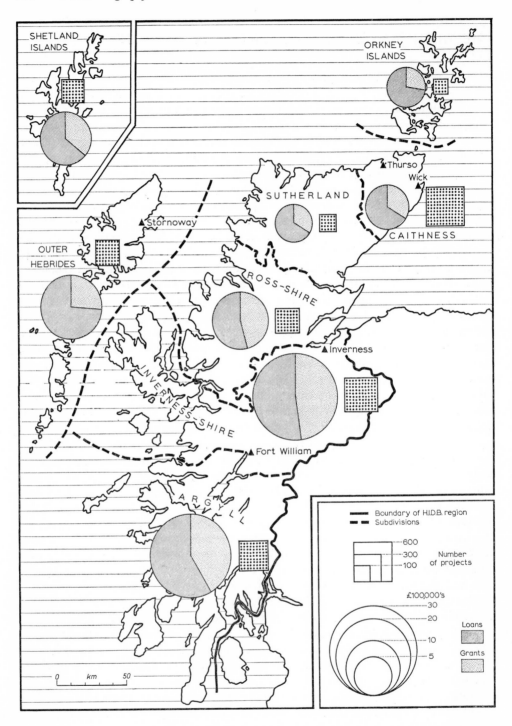

Fig. 13.1. Highlands and Islands Development Board schemes, 1965—31 March, 1972.

to enlarge farms and improve farming conditions. Opinions are divided as to whether agricultural production should continue to be subsidized or whether loans and grants should be made available only for improving farm structures and agricultural techniques. In other words, "Charity should be replaced by capital investment" (Moisley, 1964–5, p. 143).

During the present century, north-west Scotland has lost out in terms of fishing activity, which has been increasingly handled through the ports of north-east Scotland. The HIDB has operated to enlarge the volume of fish landed in the north-west and to increase the number of boats and fishing jobs in the region. The Board has received grants to bring in both new sea-going vessels and second-hand ones. Whilst it costs approximately £5000–£6000 for each new fishing job on a new vessel, average costs per job on second-hand vessels are only half that amount. Finance has also been made available for the purchase of small inshore boats of under 17 m length for lobstering, sea angling, pleasure cruising, and harbour services. The HIDB ensures that men being retrained to enter fishing receive grants during that period of readjustment.

Forestry can provide a long-term economic prop for the region, but this will not give an immediate economic return nor will it offer an instantaneous increase in employment. The northern half of Scotland contains the largest reservoir of potentially plantable land in Britain, and it is probable that the bulk of future British afforestation will take place there. The Toothill Report on Natural Resources in Scotland (1957) suggested that 300,000 ha of rough grazing land were capable of improvement by planting timber. The forestry potential has been enlarged through planting exotic species which tolerate harsh upland conditions; by cultivating the soil before planting, thereby breaking through the clay-pan layer; and by using phosphatic fertilizers to improve the soil conditions of areas to be planted. The afforested surface of north-west Scotland has almost trebled since the end of World War II and increased by 34 per cent, from 162,600 ha to 218,600 ha, between 1965 and 1970. Nevertheless, the number of Forestry Commission workers fell by 19 per cent from 2055 to 1700 over the same 5 years. Mechanization is reducing the labour involved in planting, but sawmilling and processing will employ a larger work force in the future as the Scottish forests mature.

Tourism is commonly regarded as a panacea for the ills of areas devoid of manufacturing industries. It is particularly important in planning the Highland economy since it brings an immediate financial return. In the past, the region had a marked shortage of quality accommodation. The HIDB is concerned with financing all types of accommodation schemes, from bed and breakfast in crofters' cottages, through caravan and camping sites to new hotels. It is also concerned with publicity, with investigating ways of extending the tourist season, and with ensuring that other forms of development do not jeopardize landscape quality and the chances of a further expansion of tourism in the Highlands and Islands.

The installation and expansion of manufacturing industry form important components for stabilizing employment and the resident population in the region. Only 12 per cent of the region's work force is involved in manufacturing, by contrast with 35 per cent for Scotland as a whole. The HIDB is quite clear with regard to the spatial role it may be able to play. "The task of taking jobs to every one of the pockets of unemployment is an impossible one; this is a fact which must be realistically accepted" (*First Annual Report*, 1967, p. 14). Hence the Board designated three growth areas (Caithness, Inverness/Moray Firth, and Fort William/ Lochaber) and twenty-five smaller holding points. Concentration of jobs and services in and around these centres is considered by the Board to offer the most efficient use of labour and services in an area of dispersed population to meet the needs of small new factories, services for tourism, forestry, fishing, and agriculture, and a greater concentration of costly investments on local infrastructure.

Fort William has seen the recent opening of a large pulp and paper mill. This is important to the region, first, because it performs an anchoring role for a labour force which might otherwise migrate from the region, and, second, because the mill's processing workers are maintaining larger numbers of forestry workers and their families in the Highlands. This latter figure has been estimated at between 7000 and 8000 people. The Board has recognized that clear and positive offers often have to be made rapidly if industrialists are to be attracted to install new factories. The HIDB has such powers to make definite promises for providing financial assistance and/or new buildings.

Depopulation and outmigration are at the heart of the economic and social problems of the Highlands and Islands, but suitably trained workers are sometimes not available locally when new factories are opened. To try to overcome this problem the HIDB established "Project Counterdrift". This consists of a labour register of almost 8000 (1970) workers who have left north-west Scotland but have declared themselves willing to return if suitable jobs were to become available.

The operation of the HIDB has seen an undoubted improvement in economic and social conditions in north-west Scotland. There is a growing realization that, far from being an area of disillusionment, the Highlands present opportunities for investment that place it in competition on an international basis. The psychological barriers which to date have prevented a great deal of major investment are being broken down fairly rapidly, and business men are beginning to appreciate that the Highlands and Islands offer a real alternative when they are contemplating additional plant and new capacity. Certainly the region has managed to maintain its total population numbers during the 1960's but not without significant changes in distribution, involving a retreat from the countryside (−4 per cent, 1961–70) and the further growth of the three major urban areas plus some small towns (+6 per cent).

D. Turnock (1969) envisaged that redistribution of population in the region would continue: "It can scarcely be a valid policy to resist the effects of economic and social change by underpinning all existing declining marginal areas. . . . It would be wrong to regard the present distribution of population as sacrosanct and indicative of the optimum for the Highlands; even if protection of the island population is maintained and coordinated more effectively, it seems clear that Highland development generally will be accompanied by a further substantial redistribution of population which will certainly favour the central axis [Inverness/ Fort William] and may involve the sacrifice of the most marginal communities.

This is always regrettable, but need not be taken as an admission of defeat since the scanty resources of such places can often be more economically exploited from a distance than on an intensive basis by a residential population" (p. 201).

RURAL DEVELOPMENT BOARDS

The second attempt at integrated management in parts of the British countryside stems from proposals in the 1967 Agriculture Act for the creation of rural development boards in areas of hill land. The Labour Minister of Agriculture announced that the proposed boards "were a means to help areas which have particular difficulties to face, and are to the advantage of all farmers and residents in those areas". The Act formed the culminating point of various forms of assistance which had operated in the previous two decades and attempted to encourage the agricultural industry of upland areas to adjust to the changing economic climate of farming.

The White Paper on the *Development of Agriculture* (August 1965) had emphasized that special and integrated action needed to be taken to help not only farming but also other activities in the depopulated uplands, with very low population densities, an almost complete absence of manufacturing activity, and wide stretches of unspoiled and attractive countryside. For the first time the idea of planning the uplands on a spatial or geographic basis was introduced. The White Paper paved the way for the 1967 Act which gave statutory recognition to the rural development board idea and led to specific proposals for central Wales and the northern Pennines.

The proposed boards would have particular concern for the formation of viable commercial farms; the provision of guidance in deciding whether land should be used for farming or forestry; the improvement of public services; and the preservation, but at the same time, the full use of amenities and scenery in the uplands. The proposed boards would keep all these needs and problems under review, and, in consultation with local county authorities and other bodies, would draw up proposals for the rural management of their operational areas.

The Agriculture Act made special financial grants available for improving communications and public services in the areas to be covered by the proposed boards; for installing electricity, piped water, and gas in full-time farming and forestry dwellings and in those to be used to accommodate visitors; and, also, for building up tourist accommodation facilities within the operational areas. The proposed boards would be given special powers to avoid the frustration of their programmes. Such powers included the requirement to obtain the boards' consent for schemes for the transfer of agricultural, wood- and waste-land (except among direct members of the family, and excluding residential property); and for the conversion to timber of more than 4 ha of land on a single ownership unit during any twelve-month period. Afforestation programmes would be decided after discussion between the Forestry Commission, the county planning authorities, and the proposed boards. The boards might withhold their consent from proposals for land transfer or land-use change if landowners intended to subdivide existing farms into smaller units or if the board decided that it would be better to use the farm being transferred for the rational enlargement of surrounding properties. In certain circumstances, the proposed boards would be able to acquire land compulsorily if the land in question were deemed to be essential for farm enlargement or for farm-boundary adjustment schemes. In addition, the boards could compulsorily purchase any portions of land that were transferred illegally without their prior consent. In fact, such powers of compulsory purchase were surrounded by safeguards for landowners. In addition, the 1967 Act ensured that public opinion should be thoroughly sounded out before rural development boards could be set up.

EXPERIENCE IN CENTRAL WALES

The problems of this upland region have been studied in a series of official reports since World War II. The following common features emerge from these investigations. Depopulation, in response to the absence of a stable economic base, has reduced the region's total number of residents by 25 per cent from the peak in 1871. Central Wales still has too many farms that are incapable of yielding a satisfactory living on a full-term basis. Farming needs to be modernized and the regional economy diversified through the further development of tourism and forestry as well as manufacturing, which has already been considered in Chapter 11. Public services need to be improved and rationalized.

The projected board's operational area was defined to cover 336,700 ha (Fig. 13.2). About one-eighth of the total area was owned by forestry interests. The greater part of the area consisted of poor quality land, suitable only for extensive livestock rearing, but there were some pockets of better land for growing sale crops, and rather larger areas well suited to rearing fat sheep and dairy cattle. Only 1000 of the 5560 farm units (18 per cent) in the area were commercial full-time farms. Only 200 (3.6 per cent) employed more than two men apiece. The proposals for the rural development board were advertised in September and October 1967 and then, in accordance with the Agriculture Act, a public inquiry was started at Aberystwyth in both Welsh and English to allow the opinions of local farmers and landowners to be known and recorded. In general terms, the county councils of mid Wales were in favour of the proposed board, but many local landowners were in opposition.

Their main arguments included the following points. First, some maintained that rural development boards were quite unnecessary for the management of upland areas and that the previous "free market" system was more desirable. Second, it was argued that in any case the members of the proposed boards should be elected democratically by the local population and not appointed by the Minister of Agriculture. Third, some landowners insisted that property amalgamations and farm enlargements were already proceeding rapidly without the intervention of the development board. They wanted to ensure that any proposed boards should not be granted powers of compulsory purchase. Fourth, the argument was raised that many farmers in the region were already obtaining satisfactory incomes. Fifth, in any event the landowners insisted that the financial benefits which the proposed board might bring (in the form of special grants)

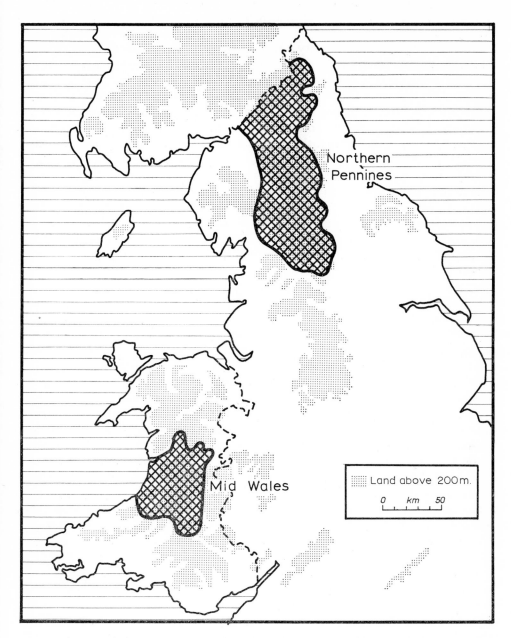

Fig. 13.2. Operational areas for the proposed rural development boards.

would be quite outweighed by the loss of the landowners' freedom to dispose of land when and how they wished, without having to seek permission from a management authority.

After the lengthy public inquiry it was clear that if the proposed rural development board were ever to be created it would need, first, to break down the hostile attitudes of the local landowners that had been expressed so freely, and, second, to find a way of winning the active co-operation of the farming community. Unfortunately these tasks were not to be accomplished. On 6 July 1970 the Secretary of State for Wales announced that the Conservative Government had decided "not to proceed with the proposal for a Rural Development Board in Wales. We believe that it would be wrong to ignore the force and strength of local feeling on this matter. We believe, moreover, that a board could make no effective contribution to the problems of its area if it lacked essential local support." But on the same day the Minister of Agriculture announced: "We do however intend to continue with the Northern Pennines Rural Development Board which has widespread local support and which we regard as a worthwhile experiment in finding solutions for the problems of the hill areas." This was not to be the case.

EXPERIENCE IN NORTHERN PENNINES

The proposals for the NPRDB managed to pass through the critical public inquiry stage and the Board was actually set up. It operated for 17 months, reviewing and managing land use in 800,000 ha of upland territory in the "grass mountain" from the Skipton Gap northwards to the Scottish border (Fig. 13.2). About 6250 farms were found in the Board's area. Three-fifths were operating at a very poor level. They did not provide enough work individually for one full-time worker apiece, and could not hope to provide more than a bare subsistence for those who lived on them, much less allow resources to be amassed that would be needed for enlarging the properties or raising agricultural productivity. One-fifth of the holdings were at the "hardship level". The NPRDB maintained that a sensible amalgamation of farms should take place. It did not envisage the wholesale removal of all small farms, maintaining that "the quality of rural life would suffer if all farms were big farms; what is needed is a variety not only of size and type of farm, but of age, skill and outlook in the members of the village community". Similarly, the Board stated that much of the landscape beauty of the region was linked to the very existence of farming activities.

All the time an effort was made to approach the problems of the area from the point of view of human beings and not simply as an economic exercise. During the first year of its operation, 375 applications for land transfer were submitted to the NPRDB (Table 26). Of these, 360 were granted, only one transfer refused, and fourteen were still under consideration. Local opposition stemmed from the fact that the Board had the power to intervene and refuse to sanction land transfers. In fact the NPRDB encouraged intending vendors of land to seek its advice before entering into further commitments. The Board was thus able to steer many projected transfers along lines which it could sanction. In addition, the Board could acquire land by refusing to give consent to proposed transfers and then directly purchasing land from would-be vendors and using it for farm enlargement. In reality, sales of hill land were dragging, and the NPRDB was often seen as the only way of disposing of land that was no longer required by its owner. Nevertheless in December 1970 a petition was organized among local farmers against the "dictatorial" powers of the Board and its ability to intervene in land transfers.

TABLE 26. NORTHERN PENNINES RURAL DEVEL-
OPMENT BOARD
LAND TRANSFER APPLICATIONS (1 NOVEMBER
1969 TO 30 SEPTEMBER 1970)

	Number	Ha
Applications received	375	
Applications granted*	360	11,459
Applications refused	1	40

* Of which 178 (3205 ha) resulted in farm amalgamations, and 11 (2895 ha) were transfers preliminary to
afforestation.
Source: NPRDB Press Notice, 23 October 1970.

Of the 360 approvals given to land-transfer proposals during the first year of the Board's existence, 178 resulted in amalgamations of property for farm enlargement and eleven (covering 2895 ha) were transfers prior to afforestation. Timber already covered 80,000 ha, namely about 10 per cent of the Board's operational area. Afforestation provided an escape route from some of the problems of impoverished hill farming and also carried with it certain fiscal advantages (regarding income tax, capital gains tax, and estate duty). The NPRDB granted licences for afforestation to be started on 2200 ha during the first year of its existence. It expected that these areas to be afforested would be extended as a result of the approval of subsequent requests. The NPDRB attempted to achieve a balance between afforestation and agricultural management, visual amenity, the preservation of flora and fauna, and other conservation interests. It tried to weigh up economic issues against aesthetic and subjective tastes as well.

This last point was a crucial one, since some of the licences granted for afforestation were for areas inside the Yorkshire Dales National Park. This aroused great opposition, not only from residents in the immediate area and from the National Park Planning Committee, but also from the Countryside Commission, from preservationists and country lovers throughout northern England, and, indeed, from many other parts of the country. The crucial proposal for afforestation in the Upper Langstrothdale section of the National Park covered 240 ha, but it was expected that there would probably be further applications. A condition was imposed by the NPRDB that a top-class landscape architect should be employed by the afforesters to produce a forestry plan that would be aesthetically pleasing. Such a condition was complied with. Nevertheless, whilst local farmers were not

TABLE 27. APPLICATIONS FOR FINANCIAL ASSIST-
ANCE FROM THE NORTHERN PENNINES RURAL
DEVELOPMENT BOARD (1 NOVEMBER 1969 TO
30 SEPTEMBER 1970)

	Number	£
Applications received	94	
Withdrawn or rejected	34	
Approved*	18	8982
Paid*	4	686

* The 22 came in the following categories: farmhouse
accommodation, 7; telephone installation, 7; rural bus
services, 4; caravan sites, 2; mains services, 2.
Source: NPRDB Press Notice, 23 October 1970.

vociferous in their complaints, many came from interested individuals and countryside organizations. The decision to permit tree-planting in the national park
must be seen as an error in public relations which the NPRDB could well have
done without.

In addition to aspects of land-use management, the Board considered ninety-
four requests for financial assistance for improvements in communications and
public services and to help farmers and foresters to derive additional income from
catering for holidaymakers (Table 27). The NPRDB expressed its desire to see
local residents benefit in this way. It noted that the extension of the M6 motorway
northwards would greatly increase the number of visitors making recreational
demands on the northern Pennines. In the context of grants for the development
of tourist facilities, the NPRDB was not an independent planning body, and in
cases where planning permission was required from county planning offices for such
development, grants could be approved from the Board only subsequent to such
permission being obtained. Table 27 shows that relatively few applications were
received in the first year of the Board's operation. Seven grants were made to
improve visitor accommodation and an equal number for helping the installation
of telephones in particularly remote places. Many farmers in the area probably
regret that they did not make applications for financial help now that the Board's
existence has been terminated.

On 18th January 1971 the Conservative Minister of Agriculture announced
to Parliament that he planned to stop the operation of the NPRDB at the end of
March 1971. He made it clear that his decision had not been based on the way
that the Board had sought to carry out the functions laid upon it by the 1967
Agriculture Act (which had been the creation of the preceding Labour Government). He insisted that his decision stemmed from the fact that the whole concept
of a rural development board, and particularly its power over land transfer, was
alien to the philosophy of the Conservative Government. Members of the NPRDB
remarked that no other organization was being established to provide help in an
area where assistance was clearly needed.

Press reactions to the announcement of the Board's dissolution were mixed. More were in favour of the termination than were against it. Many journalists reported that the NPRDB had won little praise and a fair amount of odium during its 17 months of life. It is regrettable that the Board's controversial decisions have been publicized far more than the improvements in bus services, farm enlargement schemes, and decisions to *reject* requests for afforestation in scenically attractive areas. Rejections had involved just as large an area of land as the controversial schemes which the Board had approved.

The British are well known as a nation of pragmatists and this point has been made repeatedly by the opponents of the NPRDB who argued in favour of the various forms of "gentleman's agreement" which had operated between landowners, county planning offices, the Forestry Commission, and the national parks' authorities before the creation of the Board.

In spite of the criticisms that were raised against rural development boards, many geographers regret the dissolution of this type of organization which was conceived in a geographical fashion. The purpose of the boards had been to promote balanced development in their operational areas. Members of the Countryside Commission, whilst in opposition to the Langstrothdale decision, announced that the NPRDB would probably have done a lot of good in the future. The Country Landowners' Association did not hide its dislike of the Board's compulsory powers but stated that the difficulties facing hill areas in Britain were too grave and complex to be left entirely to the play of market forces. Similarly, the National Farmers' Union leader in Northumberland expressed his regret, being particularly concerned about the positive role that the Board had played in bringing together all interested parties, so that farming, forestry, and other commercial activities might be integrated with the needs of conservation, amenity, and recreation, and the livelihood and comfort of the people. One must repeat his fear that concern for the integrated development of agriculture, tourism, forestry, settlement management, and service provision may be dispersed again among a number of separate organizations with specific interests rather than a general concern for the whole region at heart.

CONCLUSION

Rural geography is concerned with studying a wide range of social and economic phenomena in less densely populated areas which, in visual terms, may be recognized as "countryside". With the passage of time, new and changing demands have been placed on rural resources. This trend will undoubtedly continue in the future and with ever-growing intensity. Rural structures inherited from the past will have to be remodelled and new components inserted into rural landscapes. Growing numbers of urban dwellers will make the countryside a more precious resource to compensate for city living than ever before. Management proposals will need to be presented to try to obtain the best *rapprochement* between forces that operate for the piecemeal destruction of the countryside, and others which urge for its preservation or fossilization.

It is hoped that rural geographers may be encouraged to *apply* their skills to problems of relevance in countryside management. Their particular brand of applied geography needs to be of use to the community rather than forming just another series of academically respectable investigations to be filed and forgotten.

Already the geographical contribution to rural management is being expressed in several ways. At the level of higher education, more teachers and researchers are concerned with problems encountered in the countryside. Perhaps the most satisfactory type of research is that which has been commissioned by a planning agency and will therefore be used in the practical task of rural management. Increasing numbers of graduates are entering the planning profession after having taken specialist courses to give them professional qualifications in country as well as town planning to allow them to work alongside other specialist planners. Finally, as well-informed citizens, geographers have the potential to act as useful members of the many associations that operate at both national and local levels to conserve, but not necessarily to preserve, the countryside.

REFERENCES AND FURTHER READING

Rural planning in Britain and the need for an integrated, developmental approach is discussed by:

TRAVIS, A. S. (1972) Policy formulation and the planner, in ASHTON, J. and HARWOOD LONG, W. (eds.), *The Remoter Rural Areas of Britain*, Oliver & Boyd, Edinburgh, pp. 186–201.

ROBINSON, D. G. (1972) Comprehensive development, in ASHTON, J. and HARWOOD LONG, W. (eds.), *op. cit.*, pp. 215–24.

General problems of north-west Scotland are considered in:

MOISLEY, H. A. (1964–5) Land use in the Scottish Highlands: the geographical background, *Advancement of Science* **21**, 141–4.

O'DELL, A. C. (1966) Highlands and Islands development, *Scottish Geographical Magazine* **82**, 8–16.

TURNOCK, D. (1969) Regional development in the Crofting counties, *Transactions of the Institute of British Geographers* **48**, 189–204.

Problems and planning objectives for the rural development board areas are presented in:

CLOUT, H. D. (1971) End of the road for North Pennines Board, *Geographical Magazine* **43**, 443.

CLOUT, H. D. (1972) L'Aménagement des hautes terres en Angleterre, *L'Information Géographique* **36**, 29–35.

MORGAN-JONES, J. (1972) Problems and objectives in rural development board areas, in ASHTON, J. and HARWOOD LONG, W. (eds.), *op cit.*, pp. 109–29.

Education facilities for country planning in Great Britain are presented in:

DAVIDSON, J. *et al.* (1971) Focus on rural planning education, *Recreation News Supplement* (Countryside Commission) **5**, 2–9.

Problems of depopulation and economic decline and the need for rational economic planning are considered in the context of the northern Pennines in:

WILLIS, K. G. (1971) *Models of Population and Income: Economic planning in rural areas*, Agricultural Adjustment Unit, University of Newcastle-upon-Tyne, Monograph no. 1.

The work of the Highlands and Islands Development Board is presented in:

CLOUT, H. D. (1968) L'Aménagement du nord-ouest écossais: Highlands and Islands Development Board, *Norois* **15,** 536–8.
GRIEVE, R. (1972) Problems and objectives in the Highlands and Islands, in ASHTON, J. and HARWOOD LONG, W. (eds.), *op. cit.*, pp. 130–45.
HIGHLANDS and ISLANDS DEVELOPMENT BOARD, (1967–), Annual Reports, Inverness.

A discussion of the scope of applied rural geography is found in:

WHEELER, P. T. (1971) Planning the drift to the towns, *Geographical Magazine* **43,** 741–2.

The problems of upland areas in Britain are discussed by:

COUNTRY LANDOWNERS ASSOCIATION (1972) *The Uplands,* London.

AUTHOR INDEX

4

ue4 4444

ue44444444 44444444444 4444444

Medhurst, F.   134
Mendras, H.   36
Mignon, C.   54, 57–59
Mitchell, G. D.   11, 20, 24, 53, 156
Morgan-Jones, J.   66
Mounfield, P.   162

Newman, J.   152

Ogle, W.   9

Pahl, R. E.   38, 39, 43, 46, 50–52, 54
Patmore, J. A.   89, 90, 166
Pinchemel, P.   9
Pinkney, D.   24
Popplestone, G.   52

Ragatz, R. L.   73, 77
Rambaud, P.   64
Ravenstein, E. G.   26
Rees, A. D.   34, 39
Robertson, I. M. L.   8, 47
Robinson, D. G.   183, 184
Ryazanov, V. S.   154
Ryle, G. B.   89, 126

Saville, J.   13, 19, 28, 64, 142
Sheppard, J. A.   16, 19
Smith, T. L.   37
Sorokin, P.   35
Steinitz, C.   135
Stroud, D. C.   44, 53

Thijsse, J.   149
Thomas, J. G.   154
Thorns, D. C.   49, 51, 53
Travis, A. S.   183, 184
Turnock, D.   188

Van Hulten, M.   149

Warriner, D.   110
Watts, H. D.   162, 163
Webber, M.   70
Weddle, A. E.   131, 132, 134
Weller, J.   82, 102
Wendel, B.   28
Wibberley, G. P.   1, 40, 43, 46, 75, 99
Williams, W. M.   35, 37
Wirth, L.   34, 38, 39
Wolpert, J.   29

Zimmerman, C.   35

GEOGRAPHICAL INDEX

SUBJECT INDEX

Mid-Wales Development Corporation 164
Mid-Wales Industrial Development Association 160–4
Migration 4, 9–16, 19–31, 35, 76, 78
Ministry of Agriculture 124, 183, 189, 190
Ministry of Defence 91, 92
Ministry of Land and Natural Resources 184
Ministry of Transport 172

Napier Commission 185
National forest parks 127
National parks 84–96, 115, 183, 193, 195
National Parks and Access to the Countryside Act 84, 89
National Parks Commission 84, 87, 89, 90, 92, 96, 133
Nature Conservancy 129, 183

Open forests 128

Part-time farming 54, 58, 156
Peak Park Planning Board 92
Place utility 29, 76
Planning Advisory Group 84, 183
Plot consolidation 106–9
Polders 149
Postal buses 172, 179
Project Counterdrift 188

Queen Elizabeth Forest Park 127

Railways 15, 19, 28, 30, 54, 62, 87, 97, 166, 167, 169
Range of a good 144

Recreation 2, 4, 31, 43, 44, 62–64, 68, 77, 79, 82–97, 99, 115, 127–30
Recreational place utility 76
Redcliffe/Maud Report 94
Remembrement 106, 107, 109
Retirement 2, 31, 38, 47, 68, 156
Royal Commission on Local Government 94
Rural development boards 183, 184, 189–95
Rural District Councils' Association 177
Rural districts 8, 47
Rural/urban continuum 37–39
Ruralization 2, 19, 57

Scott Committee 83, 86, 159, 160
Second homes 2, 39, 44, 60, 69–79
Settlement 1, 4, 20, 46, 53, 78, 110–12, 139–56, 195
Social fallow 57
Sociology, rural 2, 8, 30, 44, 46
Sports Council 129

Thresholds 21, 113, 144, 149, 155
Toothill Report 187
Town and Country Planning Acts 82–84, 99, 142, 183
Transport 4, 5, 19, 36, 43, 54, 56, 59, 156, 166–80
Transport Act 173
Trigger areas 159

Urban villages 38
Urbanization 1, 43–79

Vedel Report 118
Village colleges 142

Worker-peasants 43, 54–60